上海大学出版社

2005年上海大学博士学位论文 34

基于变域变分全时空有限元的翼型非定常流正、反命题研究

● 作 者： 陈 池

● 专 业： 流 体 力 学

● 导 师： 刘 高 联

A Dissertation Submitted to Shanghai University for the Degree of Doctor in Engineering

Unsteady Flow Analysis and Inverse Design of Airfoils Via A New Variable-Domain Variational Space-Time Finite Element Method

Candidate: Chi Chen
Supervisor: Gao-lian Liu
Major: Fluid Mechanics

Shanghai University Press
· Shanghai ·

摘 要

随着现代叶轮机快速向高速、高温、高负荷方向发展，对设计和科研人员提出了更严峻的挑战，传统的设计方法已很难满足设计需求。针对这种情况，文中首先对国内外叶轮机和翼型的反设计和优化设计方面的现状进行了综述，讨论了现有方法，特别是具有多工况设计能力的方法，各自的优点及其局限性。论文结合国家自然科学基金项目“叶轮机气动力学新一代反命题和优化设计的研究”选题，围绕翼型的非定常振荡问题及其反问题开展了以下工作：

1 对于非定常的变分问题，单纯地把时间维当作同空间一样的维来处理是完全行不通的，必然导致计算不能进行下去。由此，必须采用全时空有限元法。由于现有的时空有限元法极其复杂，很不便于使用，我们有必要提出一种新的简单的全时空有限元法。另外，有限元计算中经常需要进行梯度的计算，其计算精度将直接关系到原始变量的计算精度。为此，本文通过两个算例，采用同样的网格以算术平均、面积加权平均、面积倒数加权平均等多种计算方式计算了同一梯度，特别提出了以高次插值函数来计算梯度的方法，试图通过数值计算比较它们的精度，为后面的计算提供指导意义。采用新型全时空有限元法对一维非定常管内可压缩流动进行求解，以及对二维弯管内定常流动的非定常化求解，证明采用新型全时空有限元法确实能获得相当令人满意的结果，并可进一步推广应用到非定常的翼型振荡问题中去。

2 对定常可压缩流动下翼型反问题的变分有限元数值解法提出了一种新的推导方法，即假设翼面变分只在 y_0 方向进行。新的推导思路和方法完全能用在振荡翼型反问题的推导中。文中详细地推导了势流控制方程下定常反问题的变分泛函新形式及其在有限单元内的离散形式。对于反问题求解，这里采用了完整的变域变分有限元公式而不是简单的差商公式，以及伪非定常的处理方式，计算表明收敛速度大大加快。并通过多次计算给出伪非定常的时间步长范围。

3 详细推导了翼型振荡绕流的正、反命题的变域变分原理。以前在推导非定常反设计的变分原理时认为翼型做的是简谐振荡，振荡角度很小，本文在推导时放弃了该假设，使得翼型振荡幅度在较大范围内都适用。本文在推导时考虑了振幅与时间有关的情况。采用第2项中提到的对翼型定常反问题的新推导方法，得到了非定常反问题的新变分原理，使得采用变域变分方法的非定常反设计更完善。并用变分原理的系统性推导途径对该变分原理进行广义化，得到了非定常反设计的广义、亚广义变分原理以及广义变分原理的普遍形式。

4 对振荡翼型的非定常绕流用全时空变分有限元法进行分析，通过计算得到了尾涡面的特征。对振荡翼型绕流正命题的变分原理进行全时空变分有限元离散，并考虑了非定常Kutta条件的应用方式。最后结果证明了非定常流动情况下尾流线的起点方向的确是与翼型的某一侧平行的结论，与Giesing-Maskell模型相当一致，也进一步证明了该模型的正确性。

5　在前面对振荡翼型非定常流动正问题的基础上对非定常单工况点的反问题进行了设计计算。文中设计了一非对称的翼型,得到了与目标翼型符合得较好的结果。

6　对当前工作进行了比较详细的总结,并对下一步要开展的工作做了展望。

关键词　反设计　非定常　变域变分　时空有限元

Abstract

In recent decades, turbomachinery is increasingly required to work under higher speed, temperature and load situations. It poses an austere challenge to designers and researchers in that the conventional design can hardly satisfy these requirements. This paper starts with a status quo investigation of the topics on the inverse design and optimum design of turbomachines and airfoils home and abroad. The merits and demerits of the existing methods are discussed in detail. The topic in this dissertation is selected on the basis of the key National Natural Science Foundation item "Research on a new generation of inverse and optimum design problems in turbomachine aerodynamics and aerothermoelastodynamics". It covers the unsteady pitching problems and the single point inverse problems. Specifically they are as the following.

1 For unsteady variational problems, time dimension can never be regarded as a space dimension. Looking them as the same will lead to nothing but divergence. So, space-time finite element method should be taken in discretization. But unfortunately, the existing space-time finite element method is extremely complicated and inconvenient. In addition, gradients are frequently asked for in finite element method. Its precision is directly related to the calculational precision of

the primary variables. In consideration of this situation, it is of great importance to discuss existing ways of getting gradients. Particularly, a new way using higher order shape functions to obtain gradients is proposed. Through comparison of the precision by different ways, a supportive selection is made. On the basis of these considerations, two problems are exemplified to evaluate the new space-time FEM. The first is about the incompressible flow in a tube, the second about the steady flow in a syphon by a time-marching method. All the calculations lead to satisfactory results, proving that our new space-time finite element method can be further generalized to solve unsteady airfoil pitching problems.

2 Based on variable-domain variational finite element method, the inverse problems about airfoils under steady compressible flows are reconsidered. Previously, the variational principles for inverse problems are obtained by assuming the variational calculus is in y direction. But for the unsteady problem of the pitching airfoil, we will take a new generalized form to consider all the actual situations. Here, the variational calculus is assumed to be in the y_0 direction when the airfoil is at idle. Then, the discretized forms by variable-domain variational finite element method are obtained. The complete variable-domain variational finite element equation, instead of the simple difference quotient equation, is coupled with pseudo-unsteady treating way to solve the inverse problems. This complete formulation has greatly boosted the convergent speed. The generalization

makes the method much more complete and applicable.

3 A detail deduction of the direct and inverse problem variable-domain variational principles for pitching airfoil is followed. Previously, the principles for unsteady inverse problems are under the assumption of small pitching angle. And, it is assumed that the variable-domain variational calculus is, for the foil itself, in different direction at different attack angle. A new complete and reasonable principle is proposed in this paper. From the detail derivation of the functional, we should say that the new one has more practical significance. The principle is then further generalized to get its sub-generalized, generalized principles and the general forms of the generalized principles.

4 The unsteady pitching flow of airfoils is analyzed by variational finite element method. The wake vortex streamline is for the first time obtained numerically during the flow field analysis. The application of unsteady Kutta condition is discussed. The results prove that the outflow of the pitching airfoil is along the tangential direction of either side of the trailing edge. That agrees exactly with the Giesing-Maskell model.

5 Based on the analysis of the flow around pitching airfoils, the corresponding inverse problem is solved by the new full space-time variable-domain variational finite element method. The designed unsymmetrical airfoil agrees well with the target airfoil, demonstrating that our method is right and effective.

6 The present work is summarized and perspectives are foreseen.

Key words Inverse problem, Unsteady, Variable-domain variational calculus, space-time finite element method (FEM).

目　　录

第一章 绪 论

1.1 概述

近二十多年来，随着现代叶轮机快速向高速、高温、高负荷发展，对设计和科研人员提出了更严峻的挑战。很多情况下，人们仍然按传统方法靠经验修正叶型，并通过风洞试验和流动计算来分析其性能。这样的过程要求设计师有丰富的设计经验和专业知识，而且会耗费大量的时间、精力和财力。这种现状所以会形成，其原因可能有二：(1) 由于设计问题中含有未知边界（乃强非线性和可能不适定性的根源），故无论是其合理的提法、建模还是解法都非常困难；(2) 设计解出的叶片有时会不切实际（不能满足强度、振动、冷却和工艺等方面的实际要求），甚至不能实现（例如：叶型不封闭、叶片厚度分布不均匀或不合理、甚至为负厚度等）[1]。同时，我们还知道，叶片在运行中，不可避免地会工作在多个工况点下，这就要求提高叶轮机不仅在设计工况下，而且在变工况下的工作性能（包括效率、稳定性和可靠性）。

气动问题按其提出的形式可分为四大类[2]：正命题、反命题、杂交命题和优化命题。反命题（反问题，逆问题，设计问题，反演问题，探测问题[3]等；Inverse problem，Design problem，etc.）是给定对物型的性能要求，在若干限制条件下寻求满足要求物型的问题。已有的综述见文献[4－7]。杂交问题是介于正命题和反命题之间的一类命题，即部分给定物型，并给定物型其他部分的性能参数，要求解其余部分性能和物型参数的命题。由于杂交问题的概念比较模糊，通常人们把它归类在反设计方法中。优化命题是将分析问题同最优化方法结合起来，通过对物型的不断修正来寻求目标函数的极值，从而完成最优化设计的一类问题。

无论是优化设计或是反设计，都是寻求一种方法，把目标函数或分布和设计物型参数结合起来。所不同的是，前者是通过构建一个函数，求解其在某些约束下的极值，通常不给定流场信息。后者则是直接把物型的变化通过给定流场信息而融入分析问题中去的。从这一点看出，两者的区分并不特别严格，因为约束或目标函数可以就是给定的流场信息的具体反映。例如，一种采用修正的 Garabedian-Mcfadden 方法的多点反设计方法[8]就是一个例证。两种方法由于都是设计问题，物型是改变的，求解区域在计算过程中也是改变的（除非通过某种变换），只有通过迭代逐步逼近的方法得到满足。由于优化设计的最优化函数取极值的过程与流场分析不是直接联系在一起，势必增加流场分析次数。而且采用传统的梯度求解方法（诸如最速下降法、可行方向法、共轭梯度法等）来求极小值，虽然收敛较快，也容易求解多变量问题，但得到的却不一定是全局最优的。如果改用能得到（理论上）全局最优的随机算法（演化算法），如遗传算法（参考文献[9-10]），模拟退火算法（参考文献[11]），和人工智能方法（参考文献[12]），以及神经网络算法（如文献[13]）等，则所需的流场分析时间是极其大的，必定要求采用改进过的相关方法才能既得到全局最优，又能减少求解分析问题的次数。

把目标分布（反设计）或目标函数（优化设计）和设计物型参数结合起来通常可以采用以下三种方法：

（1）让物型参数自由变动，不作限制，这种方法主要用于反设计，因为反设计方法不大可能引入物型方面的限制。但这种方法在运用中可能会产生比较严重的后果，如物型为负厚度，不封闭或不满足结构加工要求等。

（2）让物型上的点在一定范围内自由变动，再用拟合曲线进行光顺处理，一般用样条曲线来拟合。需要设计的参数就是物型上面的离散点的坐标。另一种处理通常是限制物型表面点的连线使其导数连续。

(3) 以一些基函数来定义物型，即如下式：

$$Shape = Shape_0 + \sum_{i=1}^{n} c_i f_i \tag{1.1}$$

这里的 f_i 就是一些已知的基(或气动)函数。文献中的基函数多数是级数，它是 Lee K. D. [14] 等人提出的，因为它简便，而且可以按泰勒展开估计误差。需要优化设计的参数就是其中的系数 c_i。这种方法可以减少设计变量的个数，是值得采用的方法。但其效果不一定最好，因为基函数的个数有限，不是完备的。

1.2 优化设计方法

这里说的优化设计是基于计算流体动力学(CFD)的，流动分析程序要求反复调用。常见的流动分析程序有全速度势方程求解程序，如文献[15-18]；Euler 方程计算程序，如文献[19-22]；考虑边界层作用的黏性/无黏相互作用的计算程序，如文献[23-25]；N-S 方程求解程序，如文献[26-28]等。其中 Euler 方程计算程序是用得最多的，也最成熟。而且，文献中大多数是采用的 Jameson[29] 的方法，用有限体积法求解的。

优化的目标函数大致有以下几类：① 性能优化：通常是对升力(系数)和阻力(系数)，或它们的合成的优化。也有采用流场信息(如压力、速度分布等)的性能优化，这就和反设计方法很类似了。给定信息的主要目的是要尽量减小或消除激波的影响，推迟流动的分离。还有采用如最大做功能力的优化目标，以及局部熵增最小(主要是为了减小激波)等。② 动态及结构优化，如最小重量、固有频率、最小振幅等等。③ 多场目标，如考虑效率、振动、重量等的组合的优化目标函数。当然可以采用加权形式把性能、结构和动态性能组合起来进行整体优化，这是最直接的想法。但这种方法所用的流动分析时间是惊人的。Jameson[30]提出了用变分的方式，通过拉格朗日乘子法把各学科的优化目标同约束有机地组合在一起得到优化目标函数。这样得到的就是无约束的优化了，所用的流场分析的时间可以大大缩减。而且，该方法还很容易进行多工况点的设计。有趣的是，这种方法还用于近年来的一大热门研究领域——可变形物型研究[31]。Lagoudas 等人[32]用遗传算法设计了形状记忆合金激发的可变形机翼，考虑了热、结构和气动力的共同作用，在满足最大升阻比的目标下进行了气翼的重构。多学科耦合优化方法是如此的诱人，以至于荷兰的国家气动实验室 NLR 也决定转向该方面的研究开发，波音公司、通用[33]、NASA[34]也对这种方法极其重视。

下面我们对优化设计方法按优化手段与流场分析的结合方式进行分类说明。

(1) 直接优化方法

直接优化思想是 Hicks[35]在 1976 年提出的。它的思想很简单，就是构造一个费用(目标)函数，通过优化理论改变设计参数再计算流场，使目标函数减小。这种方法应用起来特别方便，而且考虑的设计目标可以很多，包括多工况点的设计。在流场分析方面可以用几乎任何一种流场计算方法。这种方法由于造成了纯优化问题和纯分析问题的脱节，必然导致计算效率的低下，收敛缓慢，特别是当变量多时。改进的方法之一就是采用物型表示的第 3 种方法，这样可以在一定程度上减小未知量的个数，从而分析时间也大大减小了。如 Aidala 等人[36]就是采用的这种方法。

由所采用的优化理论可以分为两种：梯度算法和随机算法。

① 梯度算法：它是把目标函数的变化与设计参数的改变通过梯度来实现的。但这种方法具有计算时间短的优点和可能得不到全局最优的缺点。

有大量文献都是采用的这种方法，如文献[37]采用的修正的可行方向法，优化了叶片重量。对透平叶片的冷却通道的形状优化，Huang, Cheng-Huang 等人[38]采用了共轭梯度法，因为该法能容易地处理多变量问题，而且收敛很快。文献[39]中采用了一种有限元-有序线性规划法（finite element-sequential linear programming），在几何和自然频率限制的情况下，最优化叶片重量。文献[40]对直升机转子叶片采用了分级优化，提高了优化效率。用梯度算法的气动优化设计有很多，可以参考文献[41]。

② 随机算法：这种方法中设计参数是在一定的范围内自由（或几乎是自由）变化的，每一次改变都得调用一次分析程序，所以它的效率要比采用梯度算法的要低。但是，由于它采用了非梯度的算法，其实用范围可望比梯度算法要宽，而且用这种方法还可以获得全局最优解。通过改进的该方法也可以与梯度算法在速度上相差不多，因此，近十年来这种算法得到了广泛的应用。

Quagliarella 等人[42]开发了一个跨声速的气动设计方法，以全速度势来求解流场。Tong[43]也建议采用遗传算法求解复杂的气动设计问题，在他的算法中还采用了专家系统。Mosetti[44]，Fan 等人[45]都采用了遗传算法来进行优化设计。近年来并行版本的遗传算法也成功地用在了机翼的优化设计方面，如文献[46]。

在这里需要着重讲一下采用修正的 Garabedian-Mcfadden[47]的多工况点反设计方法[8]。它的设计目标函数如下式所示

$$F_0\Delta y + F_1\Delta y_x + F_2\Delta y_{xx} = C_{pt} - C_{pc} \tag{1.2}$$

其中的系数 F_0、F_1、F_2 是非负常数，C_{pt}、C_{pc} 分别表示目标压力系数和计算压力系数。y 为机翼的坐标。Δy 为机翼表面位置的改变量，Δy_x 和 Δy_{xx} 分别是改变量的一、二阶导数。

该方程经有限差分离散后得到一组三对角代数方程

$$[M]\{\overline{\Delta y}\} = \{R\} \tag{1.3}$$

从而目标函数为

$$F(\overline{\Delta y}) = \frac{1}{2}\mid [M]\{\overline{\Delta y}\} - \{R\}\mid^2 \tag{1.4}$$

当引入设计限制时，可以采用罚函数方法。对多个工况点的设计时也可以采用加权的方法把目标函数组合起来。这种方法在采用流场分析程序时可以用部分收敛的结果，由此可以带来较高的收敛效率。但上述方法比较粗糙，导致形状修正可能不是很精确，所以这种方法并不具有鲁棒性。这方面的文献很多，如文献[48]就采用这种方法进行了两点跨声速翼型杂交命题设计。文献[49]用来设计了直升机转子叶片，给定的目标是表面压力分布。

(2) 基于变分理论的优化方法

这种方法也称作基于控制论的方法。它的基本思想是运用变分理论，把误差函数写成泛函的形式，通过取变分获得目标函数改变和几何形状修正之间的关系。这种方法可望获得比较高的收敛速度，而且计算也具有鲁棒性[5]。另外，变分方法还可以把多个流场结合起来，组成伴随问题进行求解，如前面的Jameson等人提出的方法。

较多文献中应用的变分法的泛函都是沿物型(一般指机翼)Γ积分计算速度 ϕ_s 和目标速度 v_s 的差值的平方，即如下式所示

$$F(y)=\oint_{\Gamma}[\phi_s(y)-v_s]^2\mathrm{d}s \tag{1.5}$$

Angrand[50]应用该泛函设计了亚音速机翼。Beux和Dervieux[51]用欧拉方程求解了同一问题。谷传纲[52]采用了另一种变分优化方法。Frank和Shubin[53]探讨了变分法应用的一个重要问题，他们认为要使计算收敛，有必要先离散分析问题后再对离散后的问题进行变分，而不是直接对连续问题进行变分。

1.3 反设计方法

由于反设计方法能较大地提高收敛速度，从而引起了人们的极大重视。反设计方法要求引入满足一定条件的流场信息，一般是给定的速度、压力、载荷和叶片厚度、环量分布。这也构成了反设计方

法的两个难点：(1) 目标分布的选择。选目标分布的原则一般是要求尽量减小激波强度和黏性影响，防止流动的分离。随便在整个物型上给定一个速度或压力分布容易导致物型太薄、负厚度、不封闭，或不满足技术要求，甚至于给定的分布根本得不到物型。所以很多文章中建议或采用杂交命题来求解，而且在给定分布时最好用优化的方法来得出，优化的目标也和前面的优化设计中所采用的优化目标相同。文献[54]对亚、跨声速的翼型表面压力分布进行了优化，得到了广泛的应用。后来 Obayashi 和 Takanashi[55] 把遗传算法引入到了给定压力的优化上面。即使这样仍然不能保证能得到比较满意的解，因为求得的物型很有可能是锯齿形的，根本不符合工程的要求，所以有必要考虑采用曲线光顺的办法，如样条曲线(面)，或要求物型表面导数连续等。这又是一个矛盾，因为所得的物型的表面流场分布就不会完全满足设计的流场分布了。(2) 跨声速设计。全速度势方程在超声速区是双曲型方程，如果在此处也给定速度分布，就可能导致方程的不适定。由此可想而知，反设计方法用在工程设计中还很漫长！

(1) 转换平面方法(Conformal mapping)

最初的反设计基本上是采用的这种方法，较早的如我国的著名工程热物理专家吴仲华先生提出的平均流线法[2]，它的思想是在考虑无黏的物型绕流情况下，物型的表面就是一条流线。这种方法简单实用，以至到现在它还在叶轮机设计中用到。

转换平面法又可以分为四类方法：流函数-势函数法；流函数法；速度图平面法；保角变换法。

① 流函数-势函数法：最早由 Stanitz[56] 用于内流设计中，这种方法在 20 世纪六七十年代用得比较多。它的基本思想是，在无黏势流中，流函数和势函数是相互正交的，而物型又是一条流线，所以经过转化后的计算区域就变成固定区域了，而不是变坐标的。求解得到流函数后就可以经过反变换得到实际的物型坐标。这种方法最大的缺点就是它不能用于三维、有黏绕流的物型设计(尽管有人用双流函

数和一势函数求解了三维情况,或可以通过缩项后求解,但考虑黏性就会很困难)。而且还会引起密度分布的不唯一性,因为当速度的 x 分量为 0 时,其 Jacobian 行列式等于 0。

② 流函数法:同上面的基本一样,但这里的 x 方向的坐标就是原来物理平面下的坐标。

③ 速度图平面法:它是 Bauer 等人[57]提出来的。由于全速度势方程是非线性的,如果能寻求一种转换把它变成线性的,则计算将很简便,由此就有了速度图平面法。其中一个坐标是速度 q,另一个坐标是液流角 θ,转换函数是:$\mathrm{d}\phi + i\dfrac{1}{\rho}\mathrm{d}\psi = qe^{-i\theta}\mathrm{d}z$,其中的 z 是复坐标,把它们分成实部和虚部求其反变换可得到物理坐标。在实际求解中还需用到保角变换,由此这种方法仍然只限于二维的计算。文献[58]中用这种方法设计了亚声速的气翼。文献[59]把超声速和亚声速区分开来求解了跨声速的叶栅。文献[60]对这种方法进行了综述。

④ 保角变换法:早在 1945 年 Lighthill[61] 就提出了这种方法,即把翼型保角地变换到一个单位圆上,对于圆的绕流我们已经比较清楚了。通过施加压力或速度分布,再通过反变换就能得到满足给定条件的翼型(见图 1.1)。这种方法后来被 Eppler[62] 发展后具有了多工况点反设计能力。Eppler 推导的是基于前后尖的翼型的反设计而言的,这与实际情况不符合。Selig 等人[63]改善了这种方法,用于了钝尾缘翼型的反设计。通过推导,他们得出下面的转换方程:

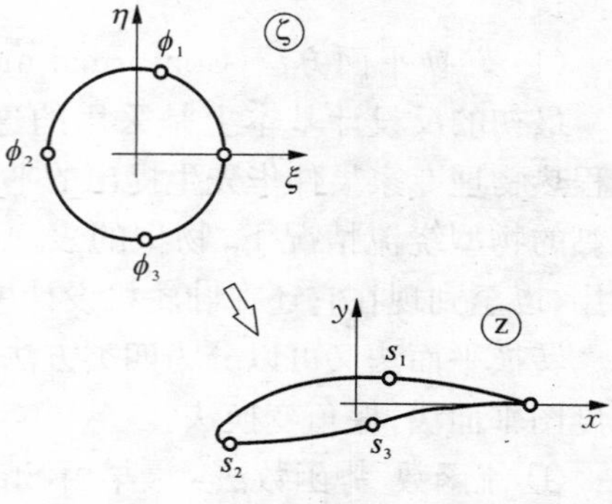

图 1.1 从圆到机翼的变换

$$x(\phi)+iy(\phi)$$
$$=-\int\left(2\sin\frac{\phi}{2}\right)^{1-\varepsilon}e^{P(\phi)}\exp\left\{i\left[\frac{\phi}{2}-\varepsilon\left(\frac{\pi}{2}-\frac{\phi}{2}\right)+Q(\phi)\right]\right\}\mathrm{d}\phi \qquad (1.6)$$

ϕ 为圆周角 $(0\leqslant\phi\leqslant 2\pi)$，$x,y$ 为机翼坐标，$P(\phi)$与给定速度分布(一般给定速度分布)与攻角相关。$Q(\phi)$是依赖 $P(\phi)$的函数。$P(\phi)$、$Q(\phi)$还必须受到一些积分限制，以保证无穷远处来流未受扰动和机翼的外形是封闭的。

反设计方法中多工况点设计能力的关键体现在给定的分布上面，即把叶片表面分成几段，对应每一段有一个相应的速度或压力分布和攻角。这也正是多工况点反设计方法最大的难点。因为翼型或叶片表面是分段的，而表面的速度分布又是给定的，有可能导致机翼表面段与段之间出现突变，这在实际情况下是不允许的，所以对给定的速度分布必须有所限制才行。当然这样的多工况点设计也可以采用多个杂交问题的设计组合来实现。

Selig 等人的文章中给出的速度分布如图 1.2 所示。最后求解所得翼型及其在多工况下的速度分布见图 1.3。可以看出这种方法在设计多工况点的翼型中还是很有效的。因为在实际设计中要求满足

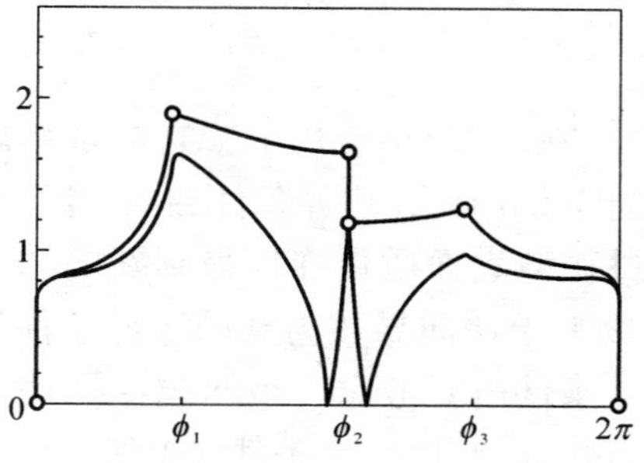

图 1.2　多(2)工况点速度分布

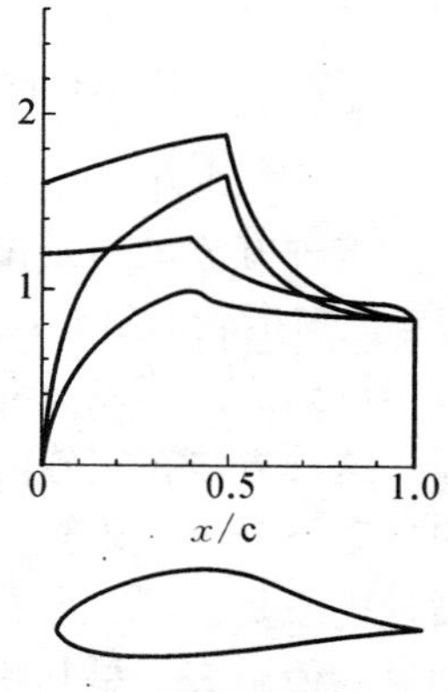

图 1.3　设计得到的翼型及其在多工况下的速度分布

结构要求，而且流体也不是无黏的，需要考虑边界层的影响，所以Selig 等人就对本方法进行了推广[64]，考虑了设计的实际要求。在文献[65]中这种方法用来设计了带吸入槽的翼型，设计结果如图 1.4 所示，展示出这种方法在设计翼型方面的较强能力。这种方法还用来设计了叶栅[66]、多单元翼型[67]、低速层流翼型[68]等。

但是这种方法有个显著的缺点：它只能用于二维的几何形状设计，不能应用到三维和非定常设计中，保角变换的性质说明了这一点。而且也不能或很难用于跨声速的机翼设计。

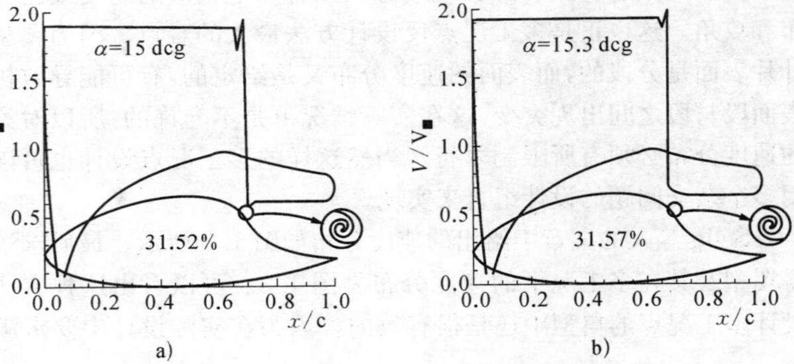

图 1.4　多点设计的 Slot-Suction 翼型(a 面元法分析的，b 设计的)

(2) 渗透方法(Transpiring)

这种方法最初是由 Henne[69]在 1980 年应用的。它的基本思想是：把物型表面目标压力分布当作 Dirichlet 边界条件加入，计算控制方程得到法向速度分量，再通过物型表面质量守恒得到新的物型表面。这种方法在实施时要求改物型上法向导数为 0 的边界条件为法向导数不为 0。要改变这种边界条件需要考虑三个问题：① 法向速度是＋或是－；② 除了静态压力外，还要加什么条件；③ 怎么计算未知量的值。表 1 是其 4 个特征值及其相容关系。文献[70]引入了三维情况下边界上的全部 5 个特征值及其兼容关系。

表 1　特征值及兼容关系

特征值	兼容关系
V_n	$-a^2(\rho^{new}-\rho^*)+(p^{new}-p^*)=0$
V_n	$V_t^{new}=V_t^*$
V_n+a	$-\rho a(V_n^{new}-V_n^*)+(p^{new}-p^*)=0$
V_n-a	$\rho a(V_n^{new}-V_n^*)+(p^{new}-p^*)=0$

V_n、V_t 分别是速度的法向、横向分量。ρ、p、a 分别是密度、压力和声速。

几何形状的修正采用了质量守恒原理，如图 1.5 所示。对图中 i 和 $i-1$ 所夹的单元运用质量守恒，得到如下的离散方程：

$$\Delta n\rho\frac{V_t|_{i-1}+V_t^{\mathrm{req}}|_{i-1}}{2}+\Delta s\frac{\rho V_n|_i+V_t^{\mathrm{req}}|_{i-1}}{2}=\Delta n\rho\frac{V_t|_i+V_t^{\mathrm{req}}|_i}{2} \tag{1.7}$$

几何修正从前驻点开始沿压力面和吸力面分开算，前驻点的 Δn 设为 0，要计算的就是其他地方的 Δn。

通过设计实验得出这种方法的效率较高。但“金无足赤，人无完人”，这种方法也有缺点：① 难以保证法向速度在收敛时趋近于 0。② 需要对分析程序进行修改。③ 较难用于多工况点的设计中。

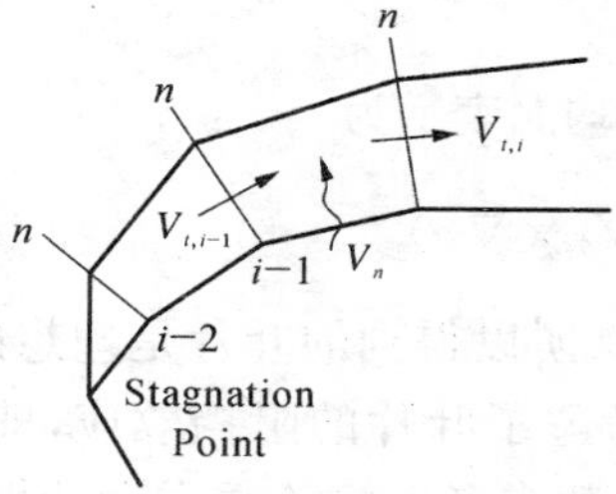

图 1.5　二维渗透原理图

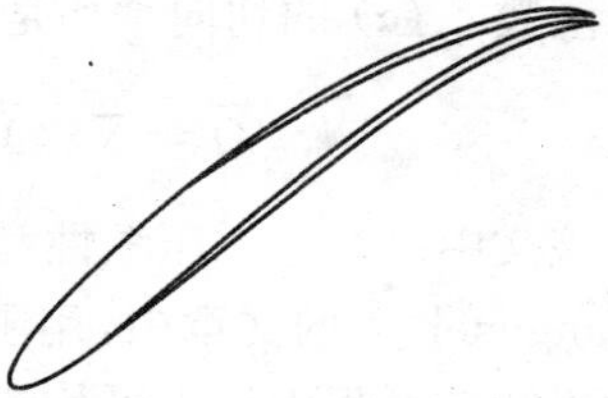

图 1.6　设计的叶片与原叶片的比较

Leonard 等人用这种方法设计了给定 Mach 数分布的亚、跨声速叶栅,得到的叶片如图 1.6 所示。而且证明本方法即使在前缘很厚的情况下也能得到比较精确的结果。Demeulenaere[71]等人把这种方法改进后用于了三维叶轮机叶片的设计中。基于 NS 方程的该方法见文献[72](二维),文献[73](三维)。

(3) 环量法(Circulation)

这种方法缘自原来的流线曲率法,其用于三维设计是 Hawthorne 等人[74]在 1984 年提出的。我们以 Borges[75]的论文来讲其基本思想。假设流动定常,无黏,进口均匀,从而流场中唯一的涡(根据 Kelvin 定理)就在叶片(0 厚度)上,可把它表示为周期三角函数形式:

$$\Omega = \nabla \times V = (\nabla r \overline{V}_\theta \times \nabla \alpha)\delta_p(\alpha) \tag{1.8}$$

其中

$$\alpha = \theta - f(r, z) = n\frac{2\pi}{B}$$

它是叶片表面变量,(所要求解的就是它,采用的是 $\overline{W} \cdot \nabla\alpha = 0$ 条件),θ 是柱坐标系统中切向坐标,$f(r, z)$ 是薄叶片表面的环形坐标,或叫翘角。$\delta_p(\alpha)$ 是周期三角函数,表示为

$$\delta_p(\alpha) = \frac{2\pi}{B}\sum_{n=-\infty}^{\infty}\delta\left(\alpha - n\frac{2\pi}{B}\right) = \mathrm{Re}\sum_{n=-\infty}^{\infty}e^{inB\alpha}$$

三角函数 $\delta_p(\alpha)$ 的切向平均是 1,因此平均涡表示为

$$\overline{\Omega} = \nabla \times \overline{V} = (\nabla r \overline{V}_\theta \times \nabla \alpha)$$

该文中没有考虑压缩性和叶片厚度,所以得到的叶片是理想的。在 Zangeneh[76]的文章中,用阻滞系数考虑了叶片的阻挡效应,即通过在连续性方程中加入平均流面厚度参数来考虑。该文中设计得到的高速(仍为亚声速的)径向入口叶轮机叶片如图 1.7 所示,结果表明在制造公差范围内这种方法是比较满意的。此后,该方法考虑了黏

性的影响[77]。在文献[78]中，它用来设计能抑制泵内二次流的混流叶轮。他们还设计了带扩散导叶的离心压缩机[79]和有分离叶片的径向透平[80]。

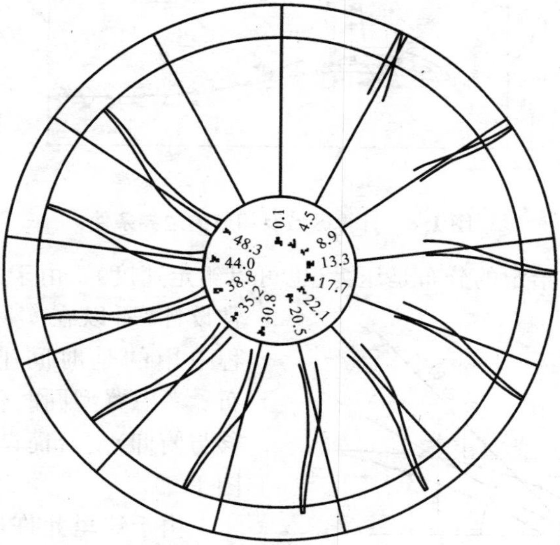

图 1.7　设计叶片的轴截面(三维)

这种方法已经比较完善了。但其给定的量是环量或速度的平均分布，与实际的环量或速度分布可能并不相同。

(4) 变域变分方法

变域变分用于气动设计是刘高联[81]在 1985 年提出的。首先是用于求解叶片间的杂交问题。后来发展后可计算有激波的跨声速流杂交问题和三维杂交问题[82-86]。在此基础上，他还发展了基于人工振荡机翼的非定常单工况点[87-89]和多工况点[90]反问题，以及基于人工振荡来流的非定常多工况点反问题[91]。

1) 单点反设计

如图 1.8 所示的计算区域内，A_1，A_2，A_3 是远场边界，A_4 是翼型的几何边界，A_f 是尾涡边界，A_s 是激波边界。

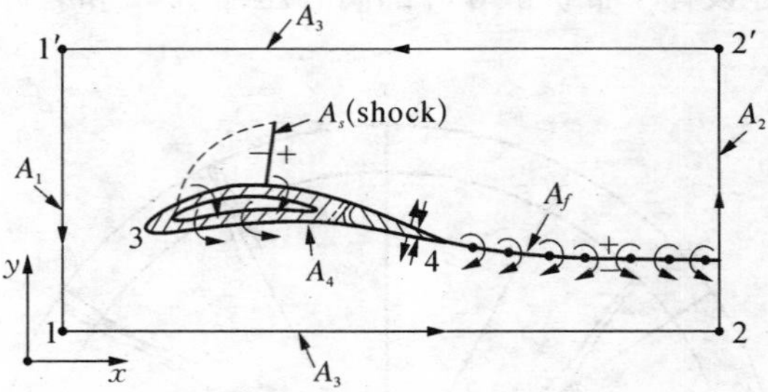

图 1.8　计算区域及相应的边界条件

在文中给定的分布是压力(也可以给定速度)。由于采用的非定常设计，所以在实际的设计过程中用的是时间平均压力分布。考虑翼型同时作弯-扭(平移与俯仰)复合简谐振动(参见图 1.9)。

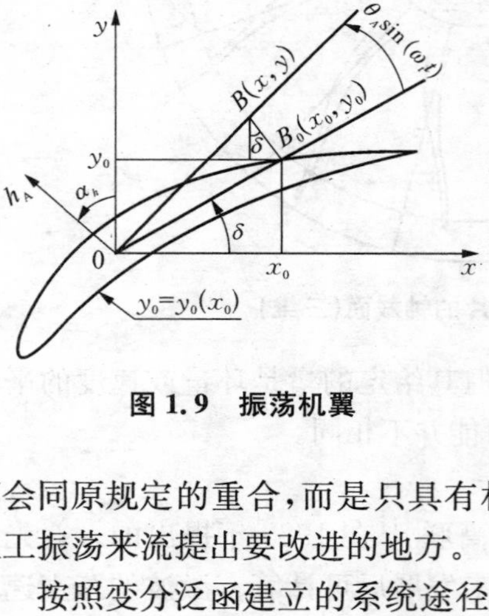

图 1.9　振荡机翼

由于这里允许规定的是翼面压力的时均分布，而不是瞬时分布。显然，同一个时均分布对应于不同的瞬时分布，因此，所解出的翼面压力分布并不会同原规定的重合，而是只具有相同的时均分布。这也是后面的人工振荡来流提出要改进的地方。

按照变分泛函建立的系统途径[92]，可以建立给定压力分布的反命题的变分泛函。上述非定常流反命题的解使下列泛函取驻值：

$$J_1(\phi, A_4, A_\Sigma, A_s, A_f) = \frac{1}{\gamma}\iiint\limits_{(V)}\left\{1-\frac{1}{m}\left[\frac{\partial\phi}{\partial t}+\frac{(\nabla\phi)^2}{2}\right]\right\}^{\gamma m} \mathrm{d}x\mathrm{d}y\mathrm{d}t + L \quad (1.9)$$

其中

$$L = \sum_{i=1}^{3} \iint_{(A_i)} (q_n)_{\mathrm{pr}} \phi \mathrm{d}A + L_b$$

$$L_b = -\frac{1}{\gamma} \oint_{(A_4)} y_0 \left\{ T(p_{\mathrm{pr}})_m - \int_0^T \overline{p \frac{\overset{\circ}{\mathrm{d}y_0}}{\mathrm{d}x_0} \theta_A \sin(2\pi f_t t) \mathrm{d}t} \right\} \mathrm{d}x_0$$

变量 ϕ、A_4、A_f、A_s 各自独立变分，表达式上有“o”的表示限制变分[93]。由此可以建立起一系列的变分原理，在此不写出了。

该方程能够满足所有的计算边界条件，远场边界用的是无反射边界条件，尾缘处采用了非定常流的广义 Kutta 条件[94]，尾涡上下面的无载荷条件。机翼表面为无滑移条件，并给定压力分布。由变域变分还可以得到该泛函满足 Rankine-Hugoniot[95]激波条件。

文献[96]用变分有限元法设计了一个翼型，得到了符合得相当好的目标压力和计算压力分布(见图 1.10)，得到的翼型与目标翼型见图 1.11。但在文中把时间和空间当作了同样的维来处理，没有用时空有限元法。本文将探讨用时空有限元法来求解非定常的反设计问题。

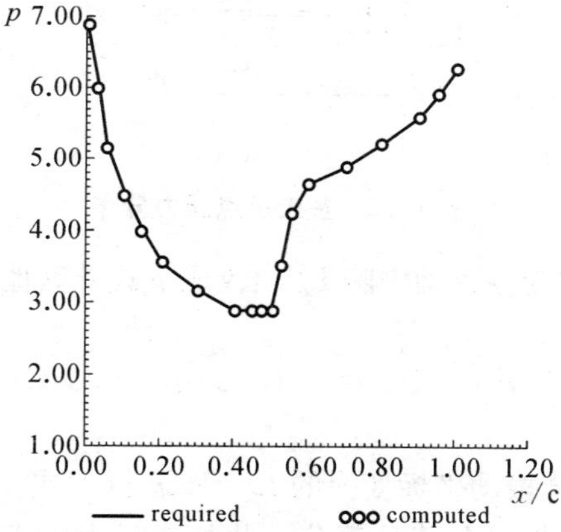

图 1.10　目标压力与计算压力分布

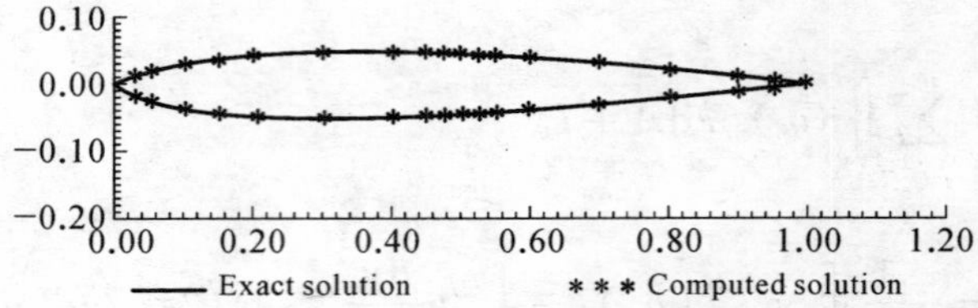

图 1.11　目标翼型与计算翼型

2）基于人工机翼振荡的多工况点设计

多点设计与前面的单点设计的主要区别在于压力的分布是根据多个工况点的压力分布在各个段上的组合。压力(系数)的给定如图 1.12 所示。

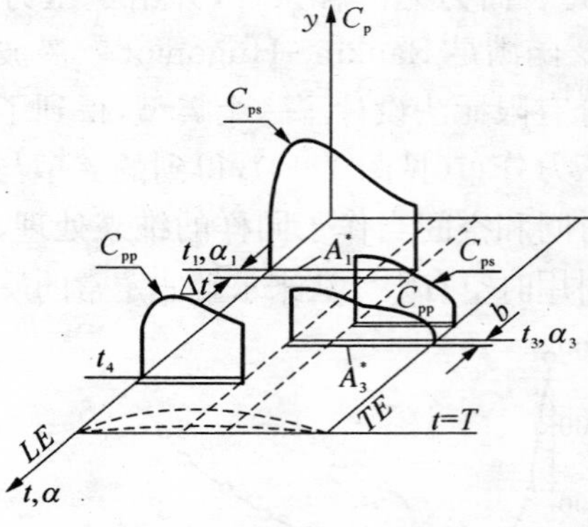

图 1.12　多工况点压力分布

最后得到的变分泛函中除 L_b 项改成下式外其他同上

$$-\frac{1}{\gamma}\int\left\{\int_{(A')}\mathring{p}\,J_a\cdot\mathring{y}_0\,\mathrm{d}t+\int_{(A_\Sigma)}p_{\mathrm{pr}}(x_0)\cdot\mathring{J}_a\cdot y_0\,\mathrm{d}t\right\}\mathrm{d}x_0 \quad (1.10)$$

其中的 J_a 为复合振动坐标变换的 $Jacobian$ 行列式。A_Σ 为施加给定压力的窄条，A' 为翼型上其余部分。此处变量上的“o”仍然表示限制变分。

3）基于人工来流振荡的多工况点设计

此时，机翼保持不动，而按实际情况对来流按不同攻角在不同时刻流入，从而可引入多工况点的设计参数。具体的原理见图 1.13，压力（系数）安排及分布同图 1.12 类似。得到的泛函变分泛函如下所示，同样，这种方法还没得到数值或分析验证。

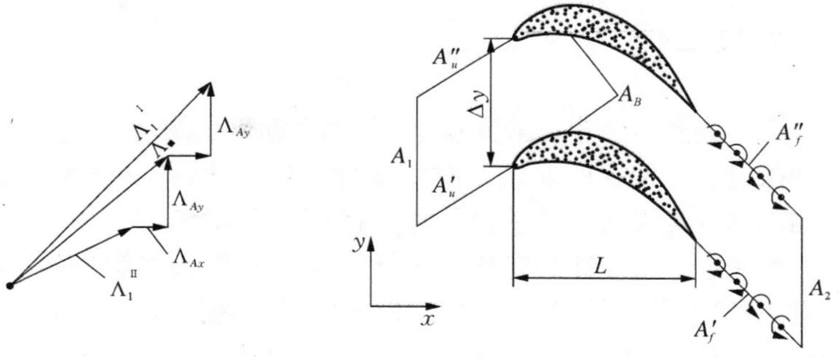

图 1.13　人工来流振荡原理图

$$J_{\mathrm{III}}(\Phi,\ A_{\Sigma},\ A_{sh},\ A_f)$$

$$=\frac{1}{\gamma}\iiint\limits_{\Omega_n}\left\{1-(\gamma-1)\left[\frac{\partial\Phi}{\partial t}+(\nabla\Phi)^2\right]\right\}^{\frac{\gamma}{\gamma-1}}\mathrm{d}x\mathrm{d}y\mathrm{d}t+L \qquad (1.11)$$

其中 Ω_n 是整个时间区域，$t_n=n\Delta t$，$(n=1,\ 2,\ 3,\cdots)$，$t_0=0$，$L=-\iint\limits_{A_{\Sigma}}\frac{1}{\gamma}p_{\mathrm{pr}}y\mathrm{d}x\mathrm{d}t+\cdots$。

（5）其他方法

拟气体方法：由于全速度势方程是混合型的，而混合型的问题很难求解，于是通过改变气体定律中的比热系数使方程保持为椭圆形的，从而可在整个区域中求解椭圆形方程。为了得到超声速区的真实解，通过求解另一初值问题，把光滑的声速线当初始条件，线上的流动变量为初值，从而可以计算得到零流线，即物型表面；修改后再计算修正的全速度势方程直到取得结果。这种方法用于了叶栅设计[97]，翼型设计[98]，进一步讨论见文献[99]。这种方法应用范围很有限(如果在超声速区不能用特征线法，则初值问题就是不适定的)。

还有其他一些方法，如面元公式法[100]。其用途也很窄，在此不提。

1.4 本文工作

本文围绕翼型的非定常流动正、反命题进行了几个方面的讨论。第二章探讨了一种新的非定常时空变分有限元法，并对一维非定常问题和二维定常流动问题进行了时空有限元法求解。第三章是推导了翼型的非定常反设计的完整的变分原理。接下来的第四章是本文的重点，分别对振荡翼型非定常绕流的正、反问题进行了计算。最后一章第五章是本文的总结以及对下一步工作的展望。

第二章　新型全时空有限元法及可压缩流动的时间相关解法

2.1　引言

非定常流动问题是当今研究的热门问题之一，但到目前为止，多数非定常问题都是用有限差分或有限体积法求解的。有限元法以其能适应复杂的几何区域而逐渐在计算流体力学领域中得到应用。但是，用有限元方法求解非定常问题的文献也主要是基于 Galerkin 变分理论的基础上进行的，对时间偏导数的处理多数是采用的差分方法。这一方面是因为非定常流动问题的变分原理的建立是极其困难的，因为非定常问题的控制方程是双曲型的，在时间方向上只能给初始条件，而不能给终止条件。为了克服这一困难，刘高联首先提出了对 Hamilton 原理的革新方法，并将它推广到双曲型问题中去，成功地建立了非定常问题的变分原理[101,102]，为非定常问题的变分求解打下了完整的理论基础。另一方面是因为差分法相对于时空有限元法要容易。但我们注意到，有很多与时间相关的问题是很难用差分法解决的，例如降落伞的降落问题，但如果用时空有限元法就容易处理得多[103]。变分法以其有强大的变域变分工具，可以促成叶轮机的多工况点反设计而逐渐得到重视。遗憾的是，迄今为止，还没有发现非定常问题变分原理有限元离散的合理方式。文献[104]对二维振荡翼型进行了变分有限元求解，但是文中对时间维的处理方式完全和空间维的处理方式等同。我们知道，非定常绝热流动的控制方程是严格的双曲型方程，对时间方向绝对不能采取同空间方向一样的处理，否则将导致计算的不稳定性。为此，我们必须采用时空有限元法，但是现有的时空有限元公式[103]极其复杂，使用很不方便，而且也不适合变分计算。为此，我们需要提出新的求解方法。受双曲型方程特征的启发，刘高联等人[105,106]提出了在处理单元内的插值函数时对时间方向上采用展开处理方式的设想。经过具体的计算试验，得到了本文的方法。

2.2 二维非定常流的控制方程及其变分理论

选滞止时的变量作无量纲量的参考量，并取长度的量纲为特征长度，我们可以推导出二维非定常跨声速流的无量纲气动方程为：

$$\frac{\partial \rho}{\partial t}+\nabla\cdot(\rho\vec{\Lambda})=0 \tag{2.1}$$

$$\nabla\phi=\vec{\Lambda} \tag{2.2}$$

$$(\gamma-1)\left(\frac{\partial\phi}{\partial t}+\frac{\Lambda^2}{2}\right)+\frac{p}{\rho}=1 \tag{2.3}$$

$$p=\rho^{\gamma} \tag{2.4}$$

其中ρ、Λ、ϕ、p、γ分别为无量纲密度、速度、位势、压力和绝热指数。

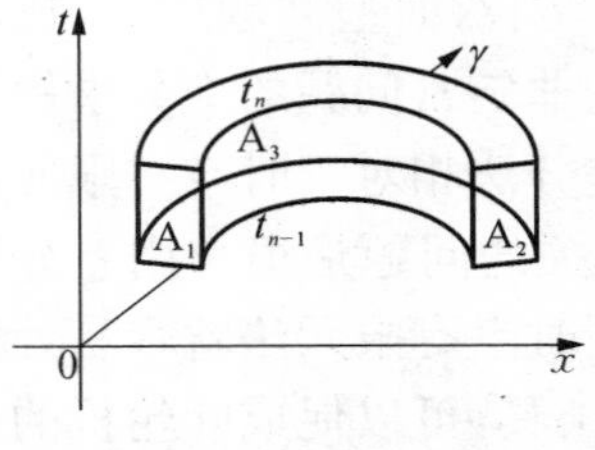

图 2.1 计算区域

如果选取如图 2.1 所示的二维管道式的计算区域，则有相应的初边值条件

(1) 在初始时刻 t_0 给定势函数 ϕ 和密度值 ρ 作为初始条件，一般稳定解与初始值无关系。具体条件为

$$\phi=f_2,\ \rho=f_3。 \tag{2.5}$$

(2) 进出口边界 A_1、A_2：应用无反射条件，即由远场数据和计算

区域内部节点得到边界上的密流

$$(\rho\Lambda_n)_{pr} = (q_n)_{pr}。\tag{2.6}$$

其中 $(\rho\Lambda_n)_{pr} = (q_n)_{pr} = \bar{\rho}(K_{\text{I}} - K_{\text{II}})/(\gamma - 1)$，$K_{\text{I}}$、$K_{\text{II}}$ 为 Riemman 不变量，其值由无穷远处及前一时间层内点值求出，$\bar{\rho}$ 为上一次近似解。

(3) 侧表面 A_3 上：由于我们这里考虑的是定常问题的时间相关解，可以假设流道是静止的，则有法向速度为零的条件，即

$$\frac{\partial\phi}{\partial n'} = 0。\tag{2.7}$$

(4) 激波面 A_s 上：如果以 U_s 表示其法向分速度，则需要满足 Rankine-Hugoniot 激波关系式

$$[|\, p\,|]/\gamma + \rho(\Lambda_n - U_s)[|\,\Lambda_n\,|] \tag{2.8a}$$

$$[|\,\rho(\Lambda_n - U_s)\,|] = 0 \tag{2.8b}$$

$$[|\,\Lambda_\tau\,|] = 0 \tag{2.8c}$$

$$[|\,H\,|] = (\gamma - 1)U_s[|\,\Lambda_n\,|] \tag{2.8d}$$

2.2.1 从 Euler 方程反推出变分原理

我们以质量守恒方程(2.1)式作 Euler 方程，进行变分泛函的反推

$$\begin{aligned}
&\iiint\left[\frac{\partial\rho}{\partial t} + \nabla\cdot(\rho\vec{\Lambda})\right]\delta\phi\,\mathrm{d}x\mathrm{d}y\mathrm{d}t \\
&= \iiint -\left[\rho\frac{\partial\delta\phi}{\partial t} + \rho\vec{\Lambda}\cdot\nabla\delta\phi\right]\mathrm{d}x\mathrm{d}y\mathrm{d}t + \\
&\iiint\left[\frac{\partial(\delta\phi\rho)}{\partial t} + \nabla\cdot(\rho\vec{\Lambda}\delta\phi)\right]\mathrm{d}x\mathrm{d}y\mathrm{d}t \\
&= -\iiint\rho\left[\delta\left(\frac{\partial\phi}{\partial t}\right) + \vec{\Lambda}\cdot\delta(\nabla\phi)\right]\mathrm{d}x\mathrm{d}y\mathrm{d}t + \delta I_b
\end{aligned}$$

$$= \frac{\iiint \rho\delta\left[1-(\gamma-1)\left(\frac{\partial\phi}{\partial t}+\frac{1}{2}\Lambda^2\right)\right]\mathrm{d}x\mathrm{d}y\mathrm{d}t}{\gamma-1}+\delta I_b$$

$$= \frac{\iiint \rho\delta\rho^{\gamma-1}\mathrm{d}x\mathrm{d}y\mathrm{d}t}{\gamma-1}+\delta I_b$$

$$= \frac{\iiint (\rho^{\gamma-1})^{\frac{1}{\gamma-1}}\delta\rho^{\gamma-1}\mathrm{d}x\mathrm{d}y\mathrm{d}t}{\gamma-1}+\delta I_b$$

$$= \frac{1}{\frac{1}{\gamma-1}+1}\frac{\iiint \delta(\rho^{\gamma-1})^{\frac{1}{\gamma-1}+1}\mathrm{d}x\mathrm{d}y\mathrm{d}t}{\gamma-1}+\delta I_b$$

$$= \frac{1}{\gamma}\delta\iiint \rho^{\gamma}\mathrm{d}x\mathrm{d}y\mathrm{d}t+\delta I_b$$

$$= \frac{1}{\gamma}\delta\iiint \left\{1-\frac{1}{m}\left[\frac{\partial\phi}{\partial t}+\frac{(\nabla\phi)^2}{2}\right]\right\}^{\gamma m}\mathrm{d}x\mathrm{d}y\mathrm{d}t+\delta I_b$$

其中 $m=\frac{1}{\gamma-1}$。从而所求泛函的主要部分为：

$$\begin{aligned} J(\phi) &= \frac{1}{\gamma}\iiint_{\Omega}\left\{1-\frac{1}{m}\left[\frac{\partial\phi}{\partial t}+\frac{(\nabla\phi)^2}{2}\right]\right\}^{\gamma m}\mathrm{d}x\mathrm{d}y\mathrm{d}t \\ &= \frac{1}{\gamma}\iiint_{\Omega} p\mathrm{d}x\mathrm{d}y\mathrm{d}t \end{aligned} \tag{2.9}$$

当我们选(2.9)式作泛函时，取变分后就应为：

$$\delta J(\phi) = \iiint_{\Omega}\left[\frac{\partial\rho}{\partial t}+\nabla\cdot(\rho\vec{\Lambda})\right]\delta\phi\,\mathrm{d}x\mathrm{d}y\mathrm{d}t-\delta I_b \tag{2.10}$$

其中 $-\delta I_b=-\iiint_{\Omega}\left[\frac{\partial(\delta\phi\rho)}{\partial t}+\nabla\cdot(\rho\vec{\Lambda}\delta\phi)\right]\mathrm{d}x\mathrm{d}y\mathrm{d}t$，也即有

$$\delta I_b = \oiint \vec{G}\cdot\vec{n'}\mathrm{d}A, \tag{2.11}$$

并已定义一个三维矢量：$\vec{G}=\rho\vec{\Lambda}\delta\phi+\rho\delta\phi\vec{i_t}=(\rho\vec{\Lambda}+\rho\vec{i_t})\delta\phi$。Ω 为内部区域，下同。$\vec{n'}$ 为图 2.1 中所示坐标下的法向矢量。$\vec{i_t}$ 为时间方向矢量。

2.2.2 边界条件的处理

从上述的计算区域图上看到，我们需要处理以下几种边界。

(1) 进出口边界 A_1、A_2

由于在边界 A_1、A_2 上我们的网格坐标与时间无关，即我们的计算区域是以流道所在平面为底，时间方向为高的一个长方体，也即是说边界变分后剩余的项式(2.11)就变为

$$-\delta I_b=-\iint_{A_1,A_2}\rho\Lambda_n\delta\phi\,\mathrm{d}s\mathrm{d}t$$

从而可以看出应在原泛函中加入的边界条件项为

$$\iint_{A_1,A_2}(\rho\Lambda_n)_{\mathrm{pr}}\phi\,\mathrm{d}s\mathrm{d}t\ \left(=\iint_{A_1,A_2}(q_n)_{\mathrm{pr}}\phi\,\mathrm{d}s\mathrm{d}t\right)\tag{2.12}$$

其中的 $(q_n)_{\mathrm{pr}}$ 是式(2.6)中的值。

(2) 物体表面边界

由于物型本身是不随时间和空间改变的，所以不需要用变域变分。而且，流道表面是不随时间改变位置的，在不添加任何项的情况下，由上面的边界剩余项式(2.11)我们立即可以得出其恰好满足 $\dfrac{\partial\phi}{\partial n'}=0$ 的条件，即式(2.7)。

(3) 激波边界 A_s

由于激波面的位置是未知的，需要应用变域变分公式。对应于该问题的边界变域变分为

$$\iint_{A_s}\left\{G'_{n'}\delta\phi+\left[F'-G_{n'}\frac{\partial\phi}{\partial n}\right]\delta e'_{n'}-G'_{n'}\frac{\partial\phi}{\partial\tau'}\delta e'_{\tau'}\right\}\mathrm{d}s\mathrm{d}t\tag{2.13}$$

其中

$F'=\frac{p}{\gamma}=\frac{1}{\gamma}\left\{1-\frac{1}{m}\left[\frac{\partial\phi}{\partial t}+\frac{(\nabla\phi)^2}{2}\right]\right\}^{\gamma m}$、$\overrightarrow{G'}=\frac{\partial F'}{\partial\phi_x}\vec{i}+\frac{\partial F'}{\partial\phi_y}\vec{j}+\frac{\partial F'}{\partial\phi_t}\overrightarrow{i_t}=-\rho(\vec{\Lambda}+\overrightarrow{i_t})$、$G'_{n'}=\overrightarrow{G'}\cdot\overrightarrow{n'}$，其中带上标“′”的都是定义在 R^3 空间中的量。

由于在激波的两侧 ϕ 值相等，即 $\phi_+=\phi_-$，即有函数 $f(x,y,t)=\phi_+-\phi_-=0$，假设其为 R^3 空间上的超曲面，显然，它在 R^2 空间中代表一个运动着的曲线(即一非定常曲线)。在此曲面上，必有

$$\mathrm{d}f=\nabla'f\cdot\overrightarrow{\mathrm{d}s'}=0 \tag{2.14}$$

其中已令

$$\begin{cases}\nabla'f=\frac{\partial f}{\partial x}\vec{i}+\frac{\partial f}{\partial y}\vec{j}+\frac{\partial f}{\partial t}\overrightarrow{i_t}\\ \overrightarrow{ds'}=\mathrm{d}x\,\vec{i}+\mathrm{d}y\,\vec{j}+\mathrm{d}t\,\overrightarrow{i_t}\end{cases}$$

即

$$|\nabla f|\,\mathrm{d}s_n+\frac{\partial f}{\partial t}\mathrm{d}t=0 \tag{2.15}$$

注意到 $\mathrm{d}s_n/\mathrm{d}t=U_s$ 正好是此曲面在 R^2 中的局部无量纲推进速度，故得

$$U_s=\frac{\mathrm{d}s_n}{\mathrm{d}t}=-\frac{\frac{\partial f}{\partial t}}{|\nabla f|} \tag{2.16}$$

而且还有

$$\left(\frac{|\nabla'f|}{|\nabla f|}\right)^2=\frac{|\nabla f|^2+\left(\frac{\partial f}{\partial t}\right)^2}{|\nabla f|^2}=1+U_s^{\,2} \tag{2.17}$$

另由 $\overrightarrow{n'}//\nabla'f$、$\vec{n}//\nabla f$，从而 $G'_{n'}$ 的表达形式可以进一步推导得出

$$G'_{n'}=\overrightarrow{G'}\cdot\overrightarrow{n'}=-\rho(\vec{\Lambda}+\vec{i_t})\left(\frac{\nabla f+\frac{\partial f}{\partial t}\vec{i_t}}{|\nabla' f|}\right)=-\frac{\rho(\Lambda_n-U_s)}{\sqrt{1+U_s^2}}。$$

同理，我们可以得到

$$\frac{\partial\phi}{\partial n'}\cdot\overrightarrow{n'}=\frac{\left(\Lambda_n-U_s\frac{\partial\phi}{\partial t}\right)}{\sqrt{1+U_s^2}} \tag{2.18}$$

由于激波是法向间断面，在其“＋”“－”表面上必定有 $\delta(\overrightarrow{e'})_+=\delta(\overrightarrow{e'})_+$，从而对激波的“＋”、“－”两侧有：

$$\iint_{A_s}\left\{(G'_{n'}\delta\phi)_+\,\mathrm{d}s+\left[\left(\frac{p}{\gamma}-G'_{n'}\frac{\partial\phi}{\partial n'}\right)\delta e'_{n'}\mathrm{d}y\right]_+-\left(G'_{n'}\frac{\partial\phi}{\partial\tau'}\delta e'_{\tau'}\right)_+\mathrm{d}s-(G'_{n'}\delta\phi)_-\,\mathrm{d}s-\left[\left(\frac{p}{\gamma}-G'_{n'}\frac{\partial\phi}{\partial n'}\right)\delta e'_{n'}\mathrm{d}y\right]_-+\left(G'_{n'}\frac{\partial\phi}{\partial\tau'}\delta e'_{\tau'}\right)_-\mathrm{d}s\right\}\mathrm{d}t \tag{2.19}$$

由 $\delta\phi_+=\delta\phi_-$，立刻可以由式(2.19)得出

$$(G'_{n'})_+=(G'_{n'})_-$$

如果引入 $[|X|]=X_+-X_-$，并把 $G'_{n'}$ 的形式代入，则有

$$[|\rho(\Lambda_n-U_s)|]=0 \tag{2.20}$$

由 $\delta x_+=\delta x_-$，可由式(2.19)导出

$$[|p|]/\gamma+\frac{\rho(\Lambda_n-U_s)}{1+U_s^2}\left[\left|\Lambda_n-U_s\frac{\partial\phi}{\partial t}\right|\right]=0$$

另外，把式(2.16)改写为

$$\left[\left|\frac{\partial\phi}{\partial t}\right|\right]=-U_s[|\Lambda_n|]$$

我们就可以把上式进一步写成

$$[|\ p\ |]/\gamma+\rho(\Lambda_n-U_s)[|\ \Lambda_n\ |]=0 \tag{2.21}$$

由 $\delta(e_{\tau'})_+=\delta(e_{\tau'})_-$ 可以导出

$$[|\ \Lambda_\tau\ |]=0 \tag{2.22}$$

把它和式(2.16)带入能量方程后得到

$$[|\ H\ |]=(\gamma-1)U_s[|\ \Lambda_n\ |] \tag{2.23}$$

其中 H 表示总焓。我们可以看出，公式(2.20)、(2.21)、(2.22)、(2.23)组合起来恰好就是 Rankine-Hugoniot 激波关系式(2.8a—2.8d)。

(4) 时间边界 t_n 和 t_{n-1}

由前面的反推过程得到的边界项，我们得到在时间方向上只有下面一项

$$-\iiint\limits_{\Omega}\frac{\partial(\rho\delta\phi)}{\partial t}\mathrm{d}x\mathrm{d}y\mathrm{d}t=-\iint\limits_{t_n}(\rho\delta\phi)_{t_n}\mathrm{d}x\mathrm{d}y+\iint\limits_{t_{n-1}}(\rho\delta\phi)_{t_{n-1}}\mathrm{d}x\mathrm{d}y$$

当 $t=t_n$ 时，上式为

$$-\iint\limits_{t_n}(\rho\delta\phi)\mathrm{d}x\mathrm{d}y$$

可以看出，当在原泛函中添加一项“限制变分”[93]，即限制密度 ρ 的变分，以上标“o”标志

$$\iint\limits_{t_n}\overset{\circ}{\rho}\,\phi\,\mathrm{d}x\mathrm{d}y \tag{2.24}$$

就可以满足时间 t_n 边界上的条件要求，并且保证了双曲型方程在 t_n 时间上不能给终值条件的要求。

当 $t=t_{n-1}$ 时：剩余项为

$$\iint\limits_{t_{n-1}}(\rho\delta\phi)\mathrm{d}x\mathrm{d}y$$

引进拉氏乘子 μ_1、μ_2 把 $\rho=\rho_{pr}$ 和 $\phi=\phi_{pr}$ 两边界条件引入

$$\begin{aligned}&\iint\limits_{t_{n-1}}\{\rho\delta\phi+\delta[\mu_1(\rho-\rho_{pr})]+\delta[\mu_2(\phi-\phi_{pr})]\}\mathrm{d}x\mathrm{d}y\\=&\iint\limits_{t_{n-1}}[\rho\delta\phi+(\rho-\rho_{pr})\delta\mu_1+\mu_1\delta\rho+(\phi-\phi_{pr})\delta\mu_2+\mu_2\delta\phi]\mathrm{d}x\mathrm{d}y\\=&\iint\limits_{t_{n-1}}[(\rho+\mu_2)\delta\phi+(\rho-\rho_{pr})\delta\mu_1+\mu_1\delta\rho+(\phi-\phi_{pr})\delta\mu_2]\mathrm{d}x\mathrm{d}y\end{aligned}$$

由于各自独立变分，从而可得

$$\mu_1=0,\mu_2=-\rho \tag{2.25}$$

由于 $\mu_1=0$，即遇到了第Ⅰ类临界变分状态，消除方法见文献[101,106]。这里可令 $\mu_1=\frac{1}{2}(\phi-\phi_{pr})$，加入到泛函中后，有

$$\begin{aligned}&\frac{1}{2}(\phi-\phi_{pr})(\rho-\rho_{pr})-\rho(\phi-\phi_{pr})\\=&\frac{1}{2}(\phi\rho-\phi\rho_{pr}-\rho\phi_{pr}+\phi_{pr}\rho_{pr})-\rho\phi+\rho\phi_{pr}\\=&\frac{1}{2}\rho\phi_{pr}-\frac{1}{2}\phi\rho-\frac{1}{2}\phi\rho_{pr}+\frac{1}{2}\phi_{pr}\rho_{pr}\end{aligned}$$

去掉常数项得

$$-\frac{1}{2}[\rho_{pr}\phi+(\phi-\phi_{pr})\rho] \tag{2.26}$$

故应在原泛函中加入的项为

$$-\frac{1}{2}\iint\limits_{t_{n-1}}[\rho_{pr}\phi+(\phi-\phi_{pr})\rho]\mathrm{d}x\mathrm{d}y \tag{2.27}$$

这是逐时间层推进时的 t_{n-1} 时刻的边界条件。对于初始时刻 t_0 来说，式中的 ϕ_{pr} 和 ρ_{pr} 就变成初始边界条件中的 f_2 和 f_3。

至此，我们把众多的边界条件都化成了自然边界条件。我们也得到了非定常可压缩流动问题的变分原理。

2.2.3 二维非定常可压流正问题的变分原理

综合上述的边界添加项，我们可以得到非定常可压缩势流的变分原理为：上述非定常可压缩势流的解将使下列泛函 J_1 取驻值：$\delta J_1=0$。其中 ϕ、A_s 各自独立变分，

$$J_1=\frac{1}{\gamma}\iiint_{\Omega}\left\{1-(\gamma-1)\left[\frac{\partial\phi}{\partial t}+\frac{1}{2}\left(\frac{\partial\phi}{\partial x}\right)^2+\frac{1}{2}\left(\frac{\partial\phi}{\partial y}\right)^2\right]\right\}^{\frac{\gamma}{\gamma-1}}\mathrm{d}x\mathrm{d}y\mathrm{d}t+$$

$$\iint_{A_1,A_2}(\rho\Lambda_n)_{\mathrm{pr}}\phi\,\mathrm{d}s\mathrm{d}t+\iint_{t_n}\overset{0}{\rho}\phi\,\mathrm{d}x\mathrm{d}y+\iint_{t_{n-1}}\{\rho(\phi-\phi_{\mathrm{pr}})-2\rho_{\mathrm{pr}}\phi\}\mathrm{d}x\mathrm{d}y。\tag{2.28}$$

值得一提的是，当计算区域中出现激波时，上述势函数控制方程已不再成立。

2.3 求解方法

2.3.1 二维泛函的离散

全时空有限元法的关键方法在于势函数在单元内的离散形式。我们提出的方法是对势函数在有限单元内作如下的离散：

$$\begin{aligned}\phi(\xi,\eta,\zeta)=&\frac{1-\xi}{2}\frac{1-\eta}{2}[\phi_1+a_1(1+\zeta)]+\\&\frac{1+\xi}{2}\frac{1-\eta}{2}[\phi_2+a_2(1+\zeta)]+\\&\frac{1+\xi}{2}\frac{1+\eta}{2}[\phi_3+a_3(1+\zeta)]+\\&\frac{1-\xi}{2}\frac{1+\eta}{2}[\phi_4+a_4(1+\zeta)]\end{aligned}\tag{2.29}$$

引入二维空间单元的形函数替代上式中含 ξ、η 的各项，即有

$$\phi(\xi,\ \eta,\ \zeta) = N_i[\phi_i + a_i(1+\zeta)]\ \text{（遵守求和约定）} \tag{2.30}$$

上两式中的 a_i 为未知数。式(2.30)中的插值函数为

$$N_1 = \frac{1-\xi}{2}\frac{1-\eta}{2},\ N_2 = \frac{1+\xi}{2}\frac{1-\eta}{2},$$

$$N_3 = \frac{1+\xi}{2}\frac{1+\eta}{2},\ N_4 = \frac{1-\xi}{2}\frac{1+\eta}{2}。$$

定义从空间-时间坐标 $(x,\ y,\ t)$ 到有限单元体积坐标 $(\xi、\eta、\zeta)$ 之间的 Jacobi 转换矩阵如下

$$J = \begin{bmatrix} \frac{\partial x}{\partial \xi} & \frac{\partial y}{\partial \xi} & \frac{\partial t}{\partial \xi} \\ \frac{\partial x}{\partial \eta} & \frac{\partial y}{\partial \eta} & \frac{\partial t}{\partial \eta} \\ \frac{\partial x}{\partial \zeta} & \frac{\partial y}{\partial \zeta} & \frac{\partial t}{\partial \zeta} \end{bmatrix} \tag{2.31}$$

从而有单元内势函数对空间-时间的偏导数

$$\begin{bmatrix} \frac{\partial \phi}{\partial x} \\ \frac{\partial \phi}{\partial y} \\ \frac{\partial \phi}{\partial t} \end{bmatrix} = J^{-1} \begin{bmatrix} \frac{\partial \phi}{\partial \xi} \\ \frac{\partial \phi}{\partial \eta} \\ \frac{\partial \phi}{\partial \zeta} \end{bmatrix} = \begin{bmatrix} J_1^{-1} \\ J_2^{-1} \\ J_3^{-1} \end{bmatrix} \begin{bmatrix} \frac{\partial \phi}{\partial \xi} \\ \frac{\partial \phi}{\partial \eta} \\ \frac{\partial \phi}{\partial \zeta} \end{bmatrix} = \begin{bmatrix} J_1^{-1} \\ J_2^{-1} \\ J_3^{-1} \end{bmatrix} \nabla'\phi \tag{2.32}$$

其中 J^{-1} 为 Jacobi 矩阵的逆矩阵，J_1^{-1}、J_2^{-1}、J_3^{-1} 分别为由 Jacobi 矩阵的逆矩阵的第一、二、三行元素组成的行向量。$\nabla'\phi$ 为势函数对体积坐标的散度，分别为

$$\frac{\partial \phi}{\partial \xi} = \frac{\partial N_i}{\partial \xi}[\phi_i + a_i(1+\zeta)] \tag{2.33a}$$

$$\frac{\partial \phi}{\partial \eta}=\frac{\partial N_i}{\partial \eta}[\phi_i+a_i(1+\zeta)] \tag{2.33b}$$

$$\frac{\partial \phi}{\partial \zeta}=N_i a_i\text{（遵守求和约定）}, i=1、2、3、4\text{（下面对 } j、k \text{ 同）。} \tag{2.33c}$$

由此我们可得到单元内泛函对任意未知数 a_i 的偏导数为

$$\frac{\partial I_e}{\partial a_i}=-\iiint_{\Omega}\rho\left[J_3^{-1}\begin{bmatrix}\frac{\partial N_i}{\partial \xi}(1+\zeta)\\ \frac{\partial N_i}{\partial \eta}(1+\zeta)\\ N_i\end{bmatrix}+J_1^{-1}\begin{bmatrix}\frac{\partial N_j}{\partial \xi}[\phi_j+a_j(1+\zeta)]\\ \frac{\partial N_j}{\partial \eta}[\phi_j+a_j(1+\zeta)]\\ N_j a_j\end{bmatrix}\cdot\right.$$

$$J_1^{-1}\begin{bmatrix}\frac{\partial N_i}{\partial \xi}(1+\zeta)\\ \frac{\partial N_i}{\partial \eta}(1+\zeta)\\ N_i\end{bmatrix}+J_2^{-1}\begin{bmatrix}\frac{\partial N_j}{\partial \xi}[\phi_j+a_j(1+\zeta)]\\ \frac{\partial N_j}{\partial \eta}[\phi_j+a_j(1+\zeta)]\\ N_j a_j\end{bmatrix}\cdot$$

$$\left.J_2^{-1}\begin{bmatrix}\frac{\partial N_i}{\partial \xi}(1+\zeta)\\ \frac{\partial N_i}{\partial \eta}(1+\zeta)\\ N_i\end{bmatrix}\right]\cdot|J|\,\mathrm{d}\xi\mathrm{d}\eta\mathrm{d}\zeta+\iint_{A_1,A_2}(\rho\Lambda_n)[N_i(1+\zeta)]\cdot$$

$$E_{13}\,\mathrm{d}\xi\mathrm{d}\zeta+\iint_{t_n}\overset{0}{\rho}N_i(1+\zeta)\cdot E_{12}\,\mathrm{d}\xi\mathrm{d}\eta \tag{2.34}$$

上列诸式中 $|J|$ 为 Jacobi 矩阵的行列式，E_{12}、E_{13} 为面积分的转换系数，其具体表示为

$$E_{12}=\sqrt{\left(\frac{\partial y}{\partial \xi}\frac{\partial t}{\partial \eta}-\frac{\partial y}{\partial \eta}\frac{\partial t}{\partial \xi}\right)^2+\left(\frac{\partial t}{\partial \xi}\frac{\partial x}{\partial \eta}-\frac{\partial t}{\partial \eta}\frac{\partial x}{\partial \xi}\right)^2+\left(\frac{\partial x}{\partial \xi}\frac{\partial y}{\partial \eta}-\frac{\partial x}{\partial \eta}\frac{\partial y}{\partial \xi}\right)^2} \tag{2.35}$$

$$E_{13}=\sqrt{\left(\frac{\partial y}{\partial \xi}\frac{\partial t}{\partial \zeta}-\frac{\partial y}{\partial \zeta}\frac{\partial t}{\partial \xi}\right)^2+\left(\frac{\partial t}{\partial \xi}\frac{\partial x}{\partial \zeta}-\frac{\partial t}{\partial \zeta}\frac{\partial x}{\partial \xi}\right)^2+\left(\frac{\partial x}{\partial \xi}\frac{\partial y}{\partial \zeta}-\frac{\partial x}{\partial \zeta}\frac{\partial y}{\partial \xi}\right)^2} \tag{2.36}$$

为了把式(2.34)中含 a 的项线性化处理，我们把 $\frac{\partial \phi}{\partial x}$ 和 $\frac{\partial \phi}{\partial y}$ 中的 a 提出来，从而可以得到单元内写成矩阵形式的代数表达式：

$$\frac{\partial I_e}{\partial a_i}=-\begin{bmatrix} b_{11} & b_{12} & b_{13} & b_{14} \\ b_{21} & b_{22} & b_{23} & b_{24} \\ b_{31} & b_{32} & b_{33} & b_{34} \\ b_{41} & b_{42} & b_{43} & b_{44} \end{bmatrix}\begin{bmatrix} a_1 \\ a_2 \\ a_3 \\ a_4 \end{bmatrix}+\begin{bmatrix} p_1 \\ p_2 \\ p_3 \\ p_4 \end{bmatrix} \tag{2.37}$$

其中的参数为

$$b_{ij}=\iiint_{\Omega}\rho\left[J_1^{-1}\begin{bmatrix}\frac{\partial N_i}{\partial \xi}(1+\zeta) \\ \frac{\partial N_i}{\partial \eta}(1+\zeta) \\ N_i\end{bmatrix}\cdot J_1^{-1}\begin{bmatrix}\frac{\partial N_j}{\partial \xi}(1+\zeta) \\ \frac{\partial N_j}{\partial \eta}(1+\zeta) \\ N_j\end{bmatrix}+\right.$$

$$\left.J_2^{-1}\begin{bmatrix}\frac{\partial N_i}{\partial \xi}(1+\zeta) \\ \frac{\partial N_i}{\partial \eta}(1+\zeta) \\ N_i\end{bmatrix}\cdot J_2^{-1}\begin{bmatrix}\frac{\partial N_j}{\partial \xi}(1+\zeta) \\ \frac{\partial N_j}{\partial \eta}(1+\zeta) \\ N_j\end{bmatrix}\right]\cdot|J|\,\mathrm{d}\xi\mathrm{d}\eta\mathrm{d}\zeta$$

$$p_i=-\iiint_{\Omega}\rho\left[J_3^{-1}\begin{bmatrix}\frac{\partial N_i}{\partial \xi}(1+\zeta) \\ \frac{\partial N_i}{\partial \eta}(1+\zeta) \\ N_i\end{bmatrix}+J_1^{-1}\begin{bmatrix}\frac{\partial N_k}{\partial \xi}\phi_k \\ \frac{\partial N_k}{\partial \eta}\phi_k \\ 0\end{bmatrix}\cdot J_1^{-1}\begin{bmatrix}\frac{\partial N_i}{\partial \xi}(1+\zeta) \\ \frac{\partial N_i}{\partial \eta}(1+\zeta) \\ N_i\end{bmatrix}+\right.$$

$$J_2^{-1}\begin{bmatrix}\dfrac{\partial N_k}{\partial \xi}\phi_k\\ \dfrac{\partial N_k}{\partial \eta}\phi_k\\ 0\end{bmatrix}\cdot J_2^{-1}\begin{bmatrix}\dfrac{\partial N_i}{\partial \xi}(1+\zeta)\\ \dfrac{\partial N_i}{\partial \eta}(1+\zeta)\\ N_i\end{bmatrix}\Bigg]\cdot|J|\,\mathrm{d}\xi\mathrm{d}\eta\mathrm{d}\zeta+$$

$$\iint_{A_1,A_2}(\rho\Lambda_n)[N_i(1+\zeta)]\cdot E_{13}\,\mathrm{d}\xi\mathrm{d}\zeta+\iint_{t_n}\overset{0}{\rho}N_i(1+\zeta)\cdot E_{12}\,\mathrm{d}\xi\mathrm{d}\eta$$

p 式中对 k 有求和约定。

合并各单元的上述表达式,并由驻值条件,就可以得到整个网格上的关于 a 的方程组

$$\frac{\partial I}{\partial a_i}=0。\tag{2.38}$$

具体计算时,初始条件取为初始密度和势函数。初始密度假设为来流的密度,初始速度大小认为在整个管道中为来流速度大小,壁面势函数就为:$\phi=\int\Lambda_\infty\mathrm{d}s$,这里的积分沿管壁面进行。因为我们的网格正交性比较好,可以取靠近壁面的内点的势函数近似与壁面点的值相等,其他点由插值决定。由于我们的计算方法是隐式的,可以取较大的时间步长。在实际计算时,为了减少计算工作量,我们采取给定某些点上的 a 值的处理方式。由此,我们以特征线(或面)公式得到进出口处的速度和密度,从而可以算出当地的 a 值。具体求解过程如下:

参照文献[107],采用 Riemann 不变量关系来处理进出口和远场无反射边界条件。进出口边界 A_1、A_2 点上的 Riemann 不变量可以写成:

$$R_\infty^-=q_{n\infty}-\frac{2c_\infty}{\gamma-1}=q_n-\frac{2c}{\gamma-1}\tag{2.39}$$

$$R_e^+=q_{ne}+\frac{2c_e}{\gamma-1}=q_n+\frac{2c}{\gamma-1}\tag{2.40}$$

上两式相加减即可得出法向速度分量和音速

$$q_n = \frac{1}{2}(R_e^+ + R_\infty^-) \tag{2.41}$$

$$c = \frac{\gamma - 1}{4}(R_e^+ - R_\infty^-) \tag{2.42}$$

一旦得到了密流 q_n 和音速 c，我们可以得到当地的速度、密度。进而由式(2.3)得到 $\frac{\partial\phi}{\partial t}$，由差分可以得到 a 的值。消去方程组式(3.38)中的已知量，减少了系数矩阵的阶次，计算时间也有所减少。

2.3.2 梯度计算方式选择

通常我们在求解某一点的梯度值时，采用的是算术平均的方法，当然也有采用有限差分法的[108-112]。但我们知道用有限差分法计算出的梯度值是节点之间的值，不是节点上的值。当计算区域比较复杂，为曲线或曲面时，用有限差分法得到的计算精度就会大打折扣。

算术平均法求某点的梯度值是用的如下公式：

$$Grad = \frac{1}{n}(grad_{(e1)} + grad_{(e2)} + \cdots + grad_{(ei)} + \cdots + grad_{(en)}) \tag{2.43}$$

其中 $grad$ 表示梯度，ei 表示第 i 个单元，n 为某点所在的单元个数。

也有推荐采用面积加权平均的，即先在各个有该点的单元中计算出该点的梯度值之后，用该单元的面积对这梯度值进行加权，再把其他单元中该点的梯度作同样的处理后，求和，再除以几个单元的面积和。如下式所示：

$$Grad = \frac{\sum grad_{(ei)} \cdot area_{(ei)}}{\sum area_{(ei)}} \tag{2.44}$$

当相邻单元面积相差不大时，两者的结果基本相同。所以，在单元划分时应避免相邻单元的面积相差太多，从而使求解的误差相近。

在实际计算中，我们还可以采用其他一些梯度计算方法，例如：

(1) 取各单元中梯度的最大值;(2) 取各单元中梯度的最小值;(3) 采用面积的倒数加权平均;(4) 采用几个单元组合成多节点单元后,取该点的梯度值。针对上面的几种梯度计算方法,我们做了一些数值实验,比较它们之间的误差,为下面的计算提供指导意义。

(1) 无限大平板间圆柱绕流

我们考虑了一个对称于两块无限大平板之间的无限长圆柱体的绕流。圆柱绕流计算在很多书上都有,也很成熟,而且还有通过映象法得到的近似解析解。由于整个区域是对称的,所以计算时我们只需要取全区域的四分之一来进行计算。

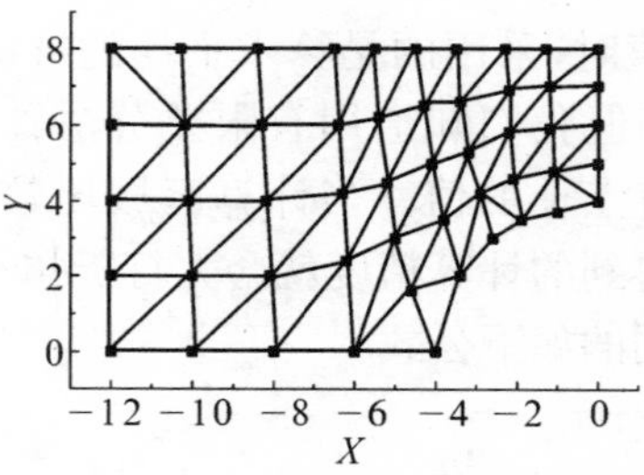

图 2.2 圆柱绕流网格

在此我们选用了两种途径求解该流场,目标都是求解区域中各点的速度。一种是先求解绕流位势,另一种是先求解绕流的流函数。两者的方程都是 Laplace 方程,计算程序除边界条件不同外基本相同。选用的单元是三角形单元,整个计算区域共 51 个节点,74 个单元,如图 2.2 所示。

该绕流流场的近似解析解是[110]:

$$V_x = \frac{\partial \psi}{\partial y} = U\left\{1 - \frac{\sinh^2(\pi b/H)\cos(2\pi y/H)}{\cosh^2(\pi x/H) - \cos^2(\pi y/H)} + \frac{\sinh^2(\pi b/H)\sin^2(2\pi y/H)}{2[\cosh^2(\pi x/H) - \cos^2(\pi y/H)]^2}\right\} \tag{2.45}$$

式中 x,y 为以圆柱中心为原点的坐标值,b 为圆柱的半径,H 为两块平板间的距离。应当指出,对于大的 b/H 比值,解析解的精度将降低(当 $b/H = 0.25$ 时,按表达式计算的误差为 2.34%)。

下面是几种计算方式的各节点(除去前驻点,因为其相对误差为无穷大)相对误差的总和:

表 1　几种计算方式 X 方向速度相对误差总和对比

计算方式	流函数解	势函数解
最大值	5.690 0	8.433 1
最小值	7.551 1	8.970 8
算术平均	2.310 7	3.740 0
面积加权平均	2.338 1	3.780 9
面积倒数加权平均	2.401 2	3.815 6

从上面的表格中可以看出，取最大、最小值的计算精度是比较差的，这是显然的，因为它们本身没什么意义。有限元计算中多数情况下采用的算术平均方式的精度是较高的，而且计算及存储量也不大。但我们也看到用面积加权平均的方法计算得到的精度也是比较好的，可以在计算中采用。原来以为面积倒数加权平均的计算精度是最高的，但实际计算后就发觉不是的。

(2) 弯管内的流动

下面采用四边形等参元网格，而且采用精确的流函数值来计算速度。计算区域及节点如图 2.3 所示。节点数 24 个，单元数 14 个。得到的结果见表 2。

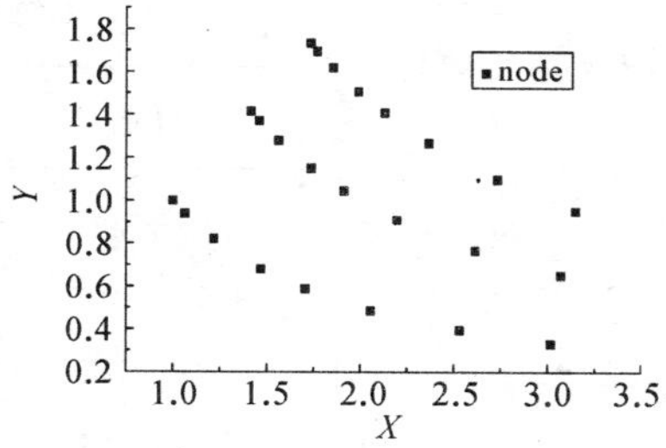

图 2.3　弯管内节点布置

表 2　几种计算方式 X 方向速度相对误差总和比较

计算方式	误差
最大	2.482 533
最小	2.297 731
算术平均	1.591 141
面积加权平均	1.649 285
面积倒数加权平均	1.573 270

本次实验中的流函数值是直接采用的解析解，但得到的速度(X方向的)精度也不是特别高，这也说明提高梯度计算的精度是很有必要的。

从表上我们可以看出这次面积倒数加权平均的误差却是最小的。算术平均的计算方式仍然显示出其精度高，存储简便的优越性，是值得采用的计算方式。同样可以看出，单纯地采用最大或最小梯度来替代节点梯度的精度是比较差的。

对于本实验，我们还考虑了采用几个单元组合成多节点单元后，取该点梯度值的方法。对于计算区域中间的6个节点，我们可以采用以它们为中心的9节点等参元来计算其X方向的速度。得到的计算结果如表3所示。

表3　节点上X方向速度值

节点号	解析解	4节点等参元		9节点等参元	
		计算值	相对误差	计算值	相对误差
10	2.918 152	2.918 390	0.000 082	2.870 284	0.016 44
11	3.127 679	3.130 398	0.000 869	3.078 874	0.015 604
12	3.473 431	3.485 547	0.003 488	3.402 593	0.020 394
13	3.823 366	3.844 016	0.005 401	3.828 978	0.001 468
14	4.394 737	4.419 260	0.022 268	4.419 599	0.005 657
15	5.244 399	5.252 843	0.001 610	5.244 399	0.002 975

可以看出以4节点等参元计算得到的梯度其精度是相当高的。另一方面也说明，梯度值计算误差较大的是边界上的一些点，因为由表2中的算术平均方式的相对误差平均后为0.065 553，但上面几点的误差都在0.023以下，而且这些点又都在区域内部。从表中我们可以得出，采用多节点等参元得到的梯度值的精度并不是很高，而且计算存储量都比较大，操作性也不强，因为原来用4节点

单元的要改为用 9 节点单元，插值函数要作变动，而且当节点处于边界上时，还得考虑用 6 节点单元等，插值函数也需要做相应的改变。

综上所述，我们认为通常选用的算术平均的梯度计算方式在多种情况下都能保证比较稳定的精度，所以我们在以后的计算中都采用这种方法。

2.4 数值结果

2.4.1 一维问题求解

为了验证前面提出的时空有限元插值模式的正确性，我们首先对一维非定常流动进行了求解[106]。选定的例子来自文献[113]，即要计算图 2.4 中当阀门打开后管内的流动状况。用本文的方法所得的结果和文献[113]用特征线所求得的结果见图 2.5 和图 2.6。可以看出两者几乎完全重合，只是在第一次波反射附近处稍微有一点不同，说明我们的方法用于计算一维非定常流动是很成功的。

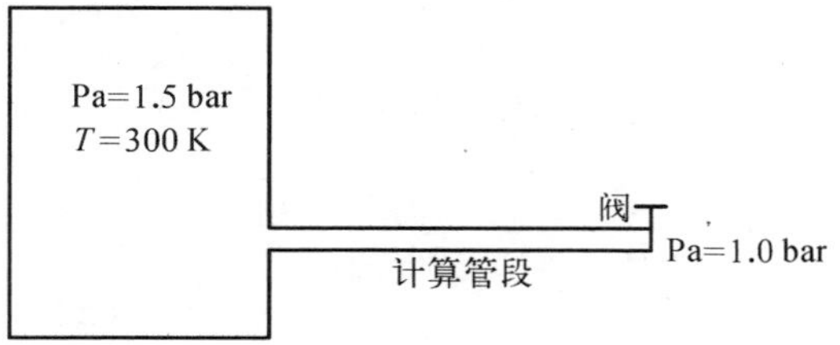

图 2.4 计算管段及相关参数

2.4.2 二维问题求解

作为上述方法的另一个应用例子，我们用时间相关法来求解定常的二维可压缩流动。计算的处理方式同一维问题求解的过程。我

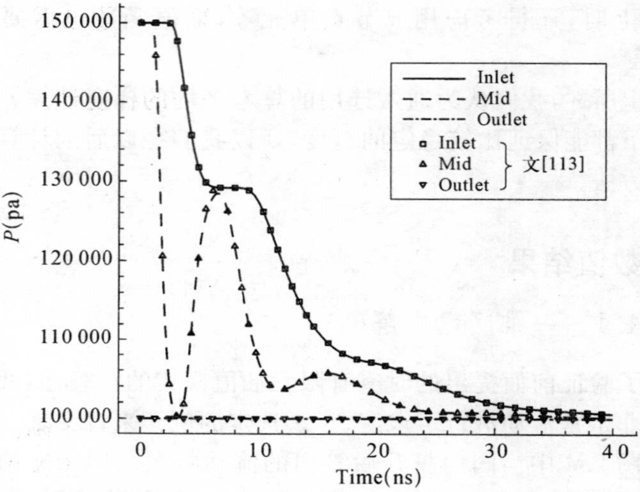

图 2.5　压力比较

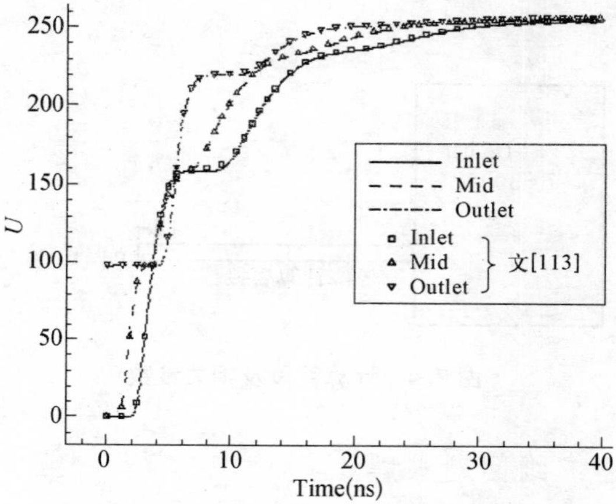

图 2.6　速度比较

们的算例选自 NACA 的研究报告，文中是用流函数和势函数坐标平面计算的。计算区域及网格如图 2.7 所示。进口无量纲速度是 0.4。我们的计算经过 12 个周期达到稳定。得到的结果(Present)和研究报告的结果(Stream-Potential)如图 2.8 和图 2.9 所示。可以看出它们是吻合得相当好的，几乎没有区别。说明我们的时间相关求解方法和有限元插值模式是很正确的。

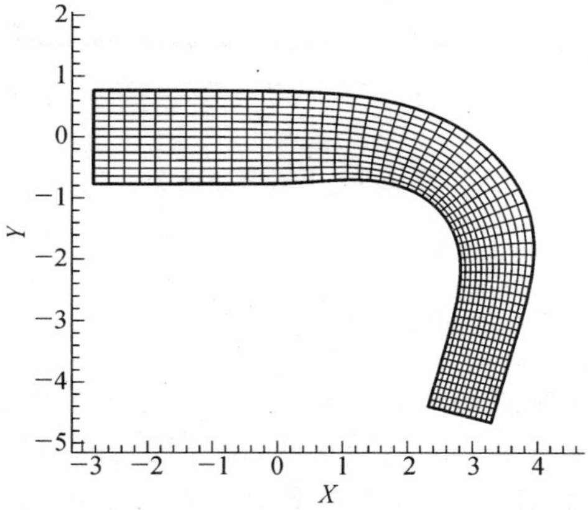

图 2.7　二维管道网格

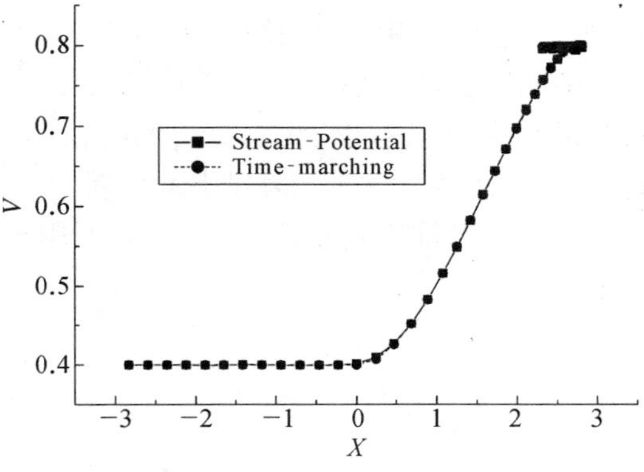

图 2.8　下管壁面速度

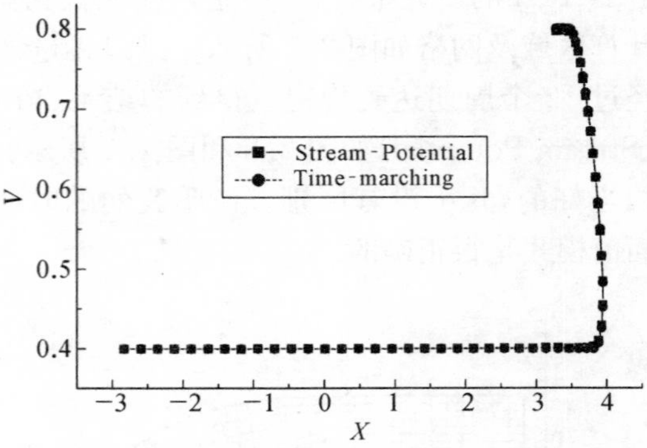

图 2.9　上管壁面速度

2.5　小结

本章在非定常流动问题的无量纲方程的基础上得到了它的变分原理。对于非定常变分问题的求解，由于现有的时空有限元法极其复杂，不适合工程计算，特别提出采用一种新型全时空有限元法来计算非定常的变分问题。由于有限元计算中梯度的计算直接关系到数值求解精度，因此文中对有限元计算中用到的几种梯度计算方式进行了精度比较，最后得出采用常规的算术平均的方式能在多数情况下取得比较稳定的精度。从文中对一维管道非定常流动和二维管道的时间相关求解的结果可以看出，我们提出的时空有限元方法和时间相关的求解方法是切实可行的。通过上面的推导过程可以看出，全时空有限元法完全可以推广到高阶的展开，并可用于高次、多维的非定常变分问题求解。

第三章　振荡翼型反命题的变分理论

3.1　问题描述

本章将研究二维振荡翼型的下列反命题：给定翼面上的瞬时压力分布，要反求出相应的翼面形状，相当于文献[90]中的 I_B 型问题。

我们已经知道非定常跨声速流动的无量纲(以滞止时的参数无量纲化)气动方程为：

$$\frac{\partial \rho}{\partial t}+\nabla \cdot (\rho \overrightarrow{\Lambda})=0 \tag{3.1}$$

$$\nabla \phi=\overrightarrow{\Lambda} \tag{3.2}$$

$$(\gamma-1)\left(\frac{\partial \phi}{\partial t}+\frac{\Lambda^2}{2}\right)+\frac{p}{\rho}=1 \tag{3.3}$$

$$p=\rho^{\gamma} \tag{3.4}$$

设翼型作俯仰振荡，如图 3.1 所示，翼型上的一点 $B_0(x_0, y_0)$ 绕振动中心 (x_1, y_1) 转过 θ 到 $B(x, y)$ 点，由图示，我们可以得到翼面瞬时坐标 $x(t)$ 和 $y(t)$ 与其翼型坐标 x_0 与 y_0 的转换关系。如果以复数来表示则有

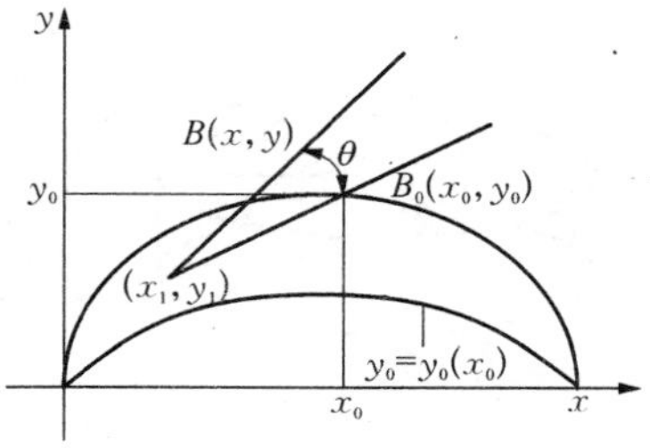

图 3.1　翼型振荡示意图

$$
\begin{aligned}
&(x+yi)-(x_1+y_1 i) \\
&=[(x_0-x_1)+(y_0-y_1)i](\cos\theta+i\sin\theta)
\end{aligned} \tag{3.5}
$$

其中的 i 为虚数单位。展开后就得到

$$x(x_0, t) = x_1 + (x_0 - x_1)\cos\theta - (y_0 - y_1)\sin\theta \tag{3.6}$$

$$y(x_0, t) = y_1 + (x_0 - x_1)\sin\theta + (y_0 - y_1)\cos\theta \tag{3.7}$$

其中 $\theta = \theta_0 + \theta_A \sin(2\pi f_t t)$，$\theta_0$ 为翼型俯仰振荡的平均攻角，θ_A 为振幅，它可以是时间 t 的函数，f_t 为振荡频率，$y_0 = y_0(x_0)$ 为翼型的坐标方程。上式是严格按照旋转公式推导的，而且这里考虑了平均攻角不为零的情况。文献[87]中的振荡公式是基于小振幅的简谐振荡情况下得出的，即认为 $\cos\theta \approx 1$，如果攻角变化范围比较大，这样的假设就有较大误差。

由式(3.6)、(3.7)可知，如果翼型的 x_0 和 t 方向不变分，则有

$$\delta x = -\delta y_0 \sin\theta,\ \delta y = \delta y_0 \cos\theta \tag{3.8}$$

我们的计算区域是围绕翼型的圆柱状区域，以翼型振荡平面为底，以时间为高。为了减少计算时对计算机存储量的要求，我们的计算分时间步进行，即把翼型振荡的一个周期分成若干个时间步，计算区域就是一个时间-空间块(叫一个 Slab)，如图 3.2 所示。在某时间层面上的计算区域如图 3.3 所示。

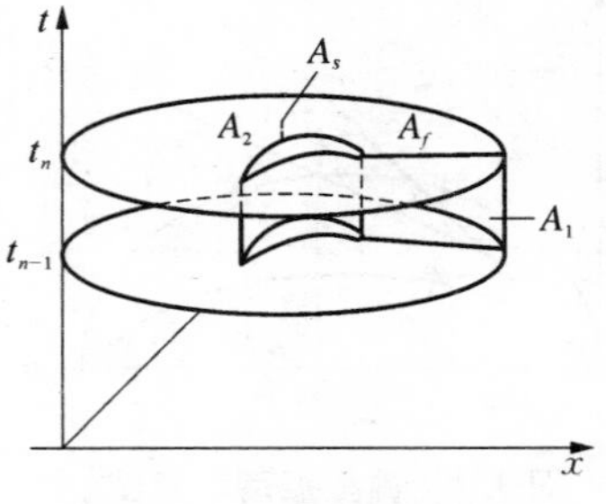

图 3.2　计算区域及边界

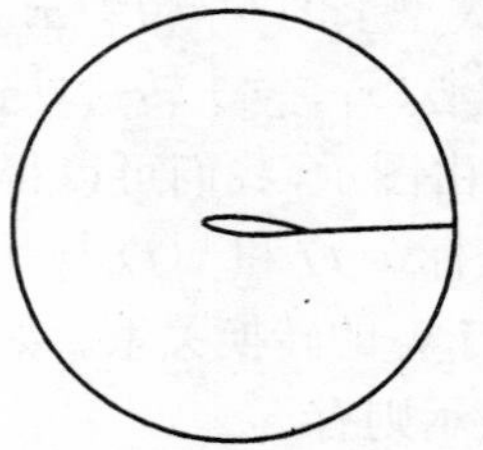

图 3.3　某时间层上的计算区域

初边值条件为

（1）在初始时刻 t_0 给定势函数 ϕ 和密度值 ρ 作为初始条件，一般初始值都是从定常问题的结果得出。具体条件为

$$\phi = f_2, \rho = f_3。 \tag{3.9}$$

（2）远场边界 A_1：应用无反射条件，即由来流数据和计算区域内部节点得到远场边界上的密流

$$(\rho\Lambda_n)_{\mathrm{pr}} = (q_n)_{\mathrm{pr}}。 \tag{3.10}$$

其中 $(\rho\Lambda_n)_{\mathrm{pr}} = (q_n)_{\mathrm{pr}} = \bar{\rho}(K_{\mathrm{I}} - K_{\mathrm{II}})/(\gamma - 1)$，$K_{\mathrm{I}}$、$K_{\mathrm{II}}$ 为 Riemman 不变量，其值由无穷远处及前一时间层内点值求出，$\bar{\rho}$ 为上一次近似解。

（3）物型面 A_2 上：由于翼型是振荡的，我们不能在每一时刻都给定目标压力分布，否则翼型就可能随时间改变。为此，我们给定某一 t_i 时刻的目标压力分布 p_{t_i}。由于泛函是整个周期积分的，从而泛函中应该给定的目标分布就为

$$p_{\mathrm{pr}} = \overset{\circ}{p} + \delta(t - t_i)(p_{t_i} - \overset{\circ}{p})。 \tag{3.11}$$

上式中的 $\delta(t - t_i)$ 为 Dirac 函数，p 上的“o”同样表示限制变分。另外，如果设翼型的法向速度为 U_2，则还有翼型表面的自然边界条件

$$\Lambda_n = U_2。 \tag{3.12}$$

（4）激波面 A_s 上：如果以 g 表示其法向分速，则有 Rankine-Hugoniot 激波关系。

$$[|\, p \,|]/\gamma + \rho(\Lambda_n - g)[|\, \Lambda_n \,|] \tag{3.13a}$$

$$[|\, \rho(\Lambda_n - g) \,|] = 0 \tag{3.13b}$$

$$[|\, \Lambda_\tau \,|] = 0 \tag{3.13c}$$

$$[|H|] = (\gamma - 1)g[|\, \Lambda_n \,|] \tag{3.13d}$$

（5）尾涡面 A_f 上下：应该满足法向速度无间断，涡面无载荷条

件。如果设尾涡面的法向速度为 U_f，这条件可以表述为

$$(\Lambda_n)_{上} = (\Lambda_n)_{下} = U_f, p_{上} = p_{下}。\tag{3.14}$$

(6) 翼型尾缘点：非定常流动的 Kutta 条件。评价 Kutta 条件的合理性是由尾缘处外边界层与无黏流交接处的流动性质所决定的，并不是真正的后缘点，因为在实际流动中，后缘处于黏性流动中，后缘点处压力差物理上自然为零。对于定常的 Kutta 条件，有以下几种表达方式：尾缘点压力连续，即尾缘载荷为零；尾缘点速度有限或者为零；无尾涡脱落等。这些方式已经被证明它们在解析的意义上是等价的。但对于非定常流动的 Kutta 条件，在提法上还有一些疑义[114-119]。文献[116,117]证实了一种新型的非定常 Kutta 条件——Giesing-Maskell 模型(简称 G－M 模型)，即：对于尾缘上游无脱体流情况，当附着涡变化时，通过尾缘点的流线与尾缘两侧切线方向一致。此外，非定常 Kutta 条件的另一种提法是：后缘点邻点(上游)压力连续，即零载荷条件。本文采用后一种 Kutta 条件提法。

3.2 翼型定常反设计新推导方式

3.2.1 变分原理

以前我们的基于变域变分的翼型反设计方法都是假设翼型的变分只在 y 坐标方向上进行。但我们可以考虑另一种翼型反设计泛函推导过程，即假设翼型的变分在某一指定的 y_0 方向进行。这对于非定常的振荡翼型，弦线与水平方向的夹角是变化的情形比较容易理解。为叙述简单以及证明方法的可行性，我们先考虑定常情况。这里我们让来流水平进入(不会失去普遍性)，而让翼型偏一个角度。由文献[81]知道，定常可压缩流动正问题的变分泛函为(问题描述见附录 B)

$$I(\phi) = \frac{1}{\gamma}\iint_{(\Omega)} \left\{1 - \frac{1}{2m}\left[\left(\frac{\partial \phi}{\partial x}\right)^2 + \left(\frac{\partial \phi}{\partial y}\right)^2\right]\right\}^{\gamma m} \mathrm{d}A + \int_{(C_{1,2,3})} (q_n)_{\mathrm{pr}} \phi \,\mathrm{d}s \tag{3.15}$$

对应于该问题的变域变分公式是

$$\delta I = \iint_{\Omega} \nabla(\rho\Lambda)\delta\phi\,\mathrm{d}x\mathrm{d}y + \int_{C_4} \frac{\partial\phi}{\partial n}\delta\phi\,\mathrm{d}s +$$

$$\int_{C_4} \left(\frac{p}{\gamma} - G_n\frac{\partial\phi}{\partial n}\right)\delta e_n\mathrm{d}s - \int_{C_4} \frac{\partial\phi}{\partial\tau}\delta e_\tau\mathrm{d}s \tag{3.16}$$

规定变分只沿着静止、水平方向的翼型的 y_0 坐标方向 $\vec{j_0}$ 上进行（见图 3.4），转动后变分的方向就变成图 3.5 中的情况了，并把 x_0 作自变量，不予变分，显然这样做丝毫不会失去普遍性。即

$$\delta\vec{e} = \delta y_0\vec{j_0} \tag{3.17}$$

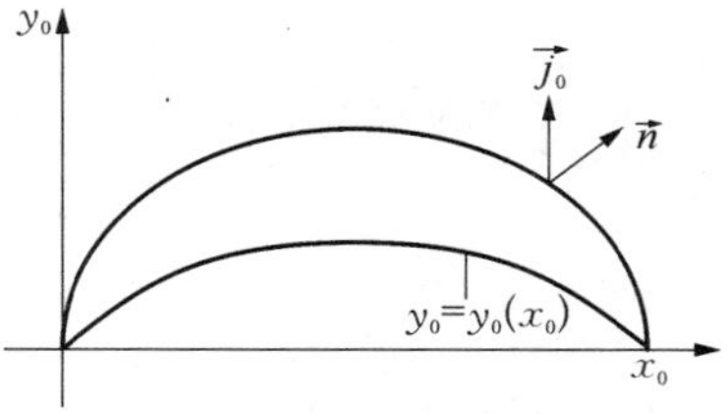

图 3.4　翼型的变分方向及其表面法向

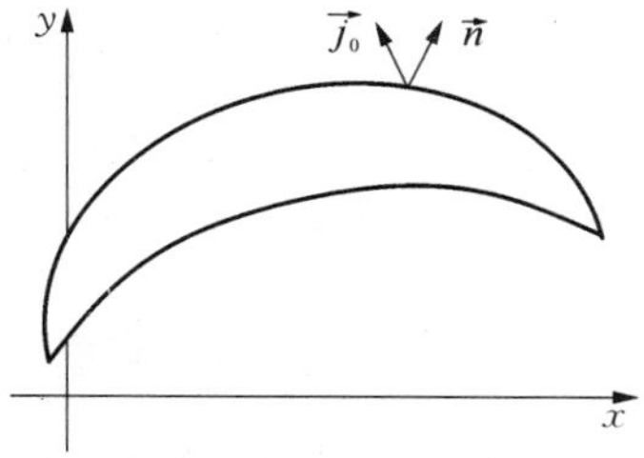

图 3.5　翼型转动一个角度后的变分方向和法向

由于翼型转动之后面积和弧长都是不变的，我们就可以把计算区域中的弧长转换成静止翼型的弧长。由于静止翼型的法向和转动后翼型的法向相对于翼型来说是完全相同的（参考图3.4、图3.5），参看图 3.6 的边界矢量关系，我们就有

$$\vec{j_0}\cdot\vec{n}\mathrm{d}s = -\mathrm{d}x_0 \tag{3.18}$$

从而有：

$$\delta\vec{e}\cdot\vec{n}\mathrm{d}s = \delta y_0\vec{j_0}\cdot\vec{n}\mathrm{d}s = -\delta y_0\mathrm{d}x_0 \tag{3.19}$$

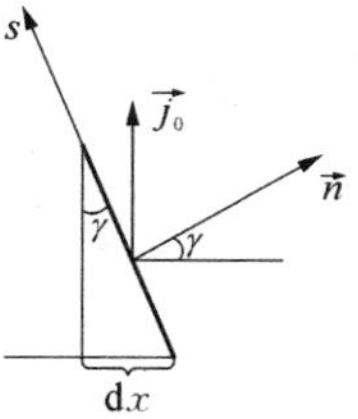

图 3.6　边界矢量关系

上面的变域变分公式中的边界项就可以写成

$$\int_{C_4} G_n\delta\phi\,\mathrm{d}s - \left[\frac{p}{\gamma} - G_n\frac{\partial\phi}{\partial n}\right]\delta y_0\,\mathrm{d}x_0 - G_n\frac{\partial\phi}{\partial\tau}\delta e_\tau\,\mathrm{d}s \tag{3.20}$$

由此不难看出，要满足翼型表面的边界条件，需要在正问题的变分泛函 I 中加入下面一项：

$$\frac{1}{\gamma}\int_{C_4} p_{\mathrm{pr}}y_0\,\mathrm{d}x_0 \tag{3.21}$$

就能使其满足待求翼型在其表面的边界条件。

综上，我们得到翼型设计的新变分泛函

$$I'(\phi,\ y_0) = \frac{1}{\gamma}\iint_{(\Omega)}\left\{1-\frac{1}{2m}\left[\left(\frac{\partial\phi}{\partial x}\right)^2+\left(\frac{\partial\phi}{\partial y}\right)^2\right]\right\}^{\gamma m}\mathrm{d}A + \int_{(C_{1,2,3})}(q_n)_{\mathrm{pr}}\phi\,\mathrm{d}s + \int_{(C_4)}\frac{p_{\mathrm{pr}}}{\gamma}y_0\,\mathrm{d}x_0 \tag{3.22}$$

3.2.2 泛函的离散

这里需要特别指出的是，泛函(3.22)中需要求的变量是 y_0，所以我们在求泛函对未知变量的导数时也应该对它求导。即

$$\begin{aligned}\frac{\partial I'_e(\phi,\ Y_{0i})}{\partial Y_{0i}} = &\iint_{(\Omega)_e}(-\rho)\left[J_1^{-1}\,\nabla_e\phi\cdot\left(\frac{\partial J_1^{-1}}{\partial x}\frac{\partial x}{\partial y_{0i}}+\frac{\partial J_1^{-1}}{\partial y}\frac{\partial y}{\partial y_{0i}}\right)+\right.\\ &\left.J_2^{-1}\,\nabla_e\phi\cdot\left(\frac{\partial J_2^{-1}}{\partial x}\frac{\partial x}{\partial y_{0i}}+\frac{\partial J_2^{-1}}{\partial y}\frac{\partial y}{\partial y_{0i}}\right)\right]|J|\,\mathrm{d}\xi\mathrm{d}\eta+\\ &\iint_{(\Omega)_e}\frac{\rho^\gamma}{\gamma}\left(\frac{\partial|J|}{\partial x}\frac{\partial x}{\partial y_0}+\frac{\partial|J|}{\partial y}\frac{\partial y}{\partial y_0}\right)\mathrm{d}\xi\mathrm{d}\eta+\\ &\int_{(C_4)_e}\frac{p_{\mathrm{pr}}}{\gamma}N_i\frac{\partial x_0}{\partial\xi}\mathrm{d}\xi\end{aligned} \tag{3.23}$$

其中

$$\phi(\xi,\ \eta) = N_i(\xi,\ \eta)\phi_i$$

$$\rho(\xi,\ \eta) = N_i(\xi,\ \eta)\rho_i, i = 1,2,3,4\ (\text{求和约定})$$

$$x_0(\xi) = F_j(\xi)x_{0j}$$

$$p_{\mathrm{pr}}(\xi) = F_j(\xi)p_{\mathrm{pr}j}, j = 1,2\ (\text{求和约定})$$

这里的 F_j 是一维有限元的型函数，$F_1 = \dfrac{1-\xi}{2}, F_2 = \dfrac{1+\xi}{2}$。

由式(3.6)、式(3.7)我们得到：

$$\frac{\partial x}{\partial y_0} = -\sin\theta \tag{3.24}$$

$$\frac{\partial y}{\partial y_0} = \cos\theta \tag{3.25}$$

上式中其他偏导数的求解过程可见相关文献(如文[120])。

3.2.3 变域变分泛函的求解

对已经求出的单元上的 $\dfrac{\partial I'_e}{\partial Y_{0i}}$ 进行总体合成，得到非线性方程组

$$\frac{\partial I'}{\partial Y_{0i}} = f_i(Y_{01},\ Y_{02}\cdots,\ Y_{0n}) = 0 \tag{3.26}$$

当然可以直接用求解非线性方程组的方法求解上式强非线性方程组，但这样做势必增加计算时间。这里我们用一种较简单的方法：伪非定常法，也就是引入 Y_0 的时间相关项，即

$$\frac{\partial Y_{0i}}{\partial t} + f_i(Y_{01},\ Y_{02}\cdots,\ Y_{0n}) = 0 \tag{3.27}$$

当时间 $t \rightarrow \infty$ 时，所得到的解为定常解，即是所要求的位置量 Y_0 的值。这个方程可以采用简单的向前差的方式，如下

$$Y_{0i}^{n+1} = Y_{0i}^{n} - \Delta t \times f_i(Y_{01},\ Y_{02}\cdots,\ Y_{0n}) \tag{3.28}$$

上式中 Y_{0i}^{n+1} 是将要计算的第 i 点的 Y_0 值，而 Y_{0i}^{n} 是上一步计算得到的第 i 点的 Y_0 值。在数值计算中，发现 Δt 的取法与解的精度和稳定性有关，计算发现取 $\Delta t=0.5—1.5$ 比较合适，本文取 1.0。

3.2.4 设计算例

我们这里考虑两个有攻角的设计算例，其中一个是用 NACA0012 来设计带 2°攻角的 NACA2412 翼型。我们的初始网格是 NACA0012 的有 2°攻角的网格，如图 3.7 所示，可以看出这里的翼型是倾斜放置的。来流 Mach 数定为 0.5。计算经过 700 次正计算收敛到预定的要求。设计的翼型结果见图 3.8，压力比较见图 3.9。压力系数等值线见图 3.10。从图 3.8 和 3.9 都可看出我们设计出的翼型和目标翼型吻合得相当好，只是在前后缘附近才有一些差别。

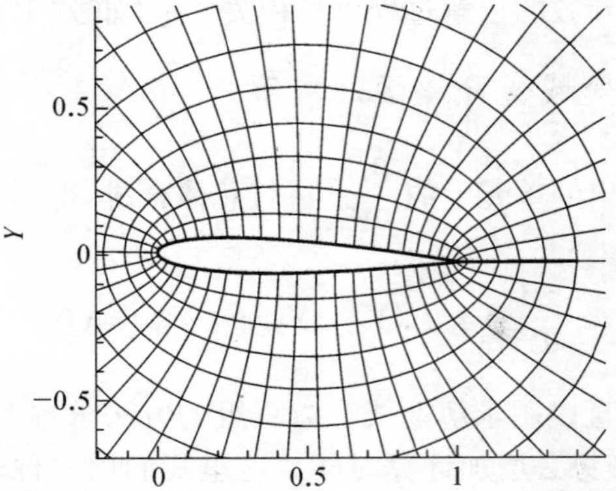

图 3.7 NACA0012 在 2°攻角时的网格

另一个算例是设计有 4°攻角的 NACA0012，初始翼型为 NACA2412。由于我们的计算还不能考虑激波，所以我们把来流 Mach 数定为 0.3。从设计得到的翼型结果（见图 3.11）和压力系数图（图 3.12）都能说明我们的设计是很成功的，即使是在攻角比较大的情况下都行。

通过这两个例子说明我们的公式，我们的方法是完全正确的，为下面的振荡翼型反设计公式的得出有重要意义。

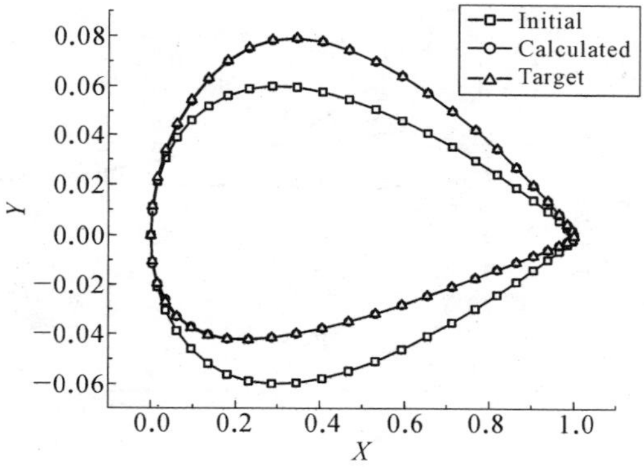

图 3.8　有 2°攻角的 NACA2412 重构

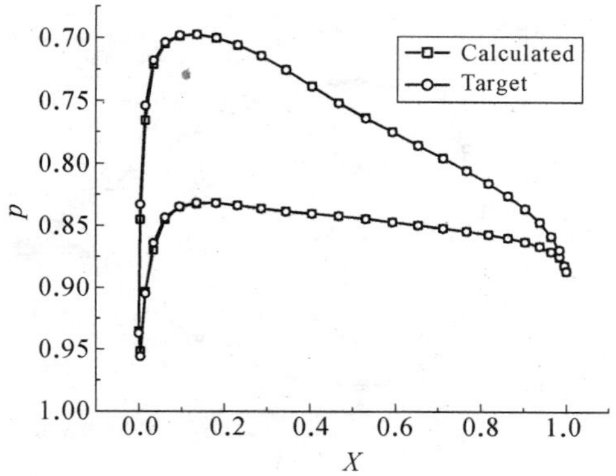

图 3.9　有 2°攻角的 NACA2412 的压力对比

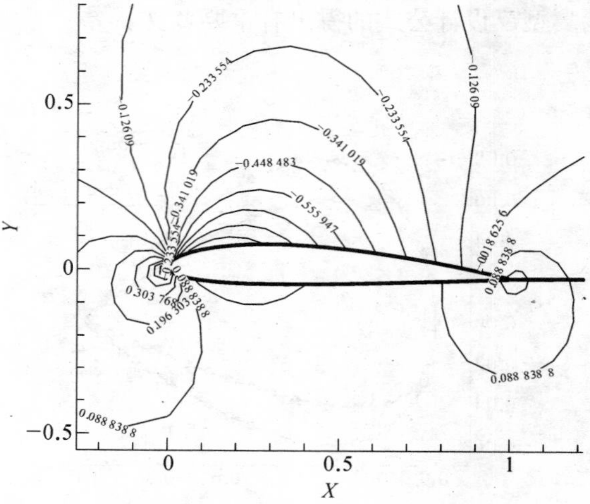

图 3.10　NACA2412 压力系数等值线

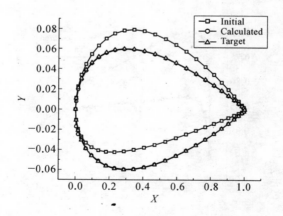

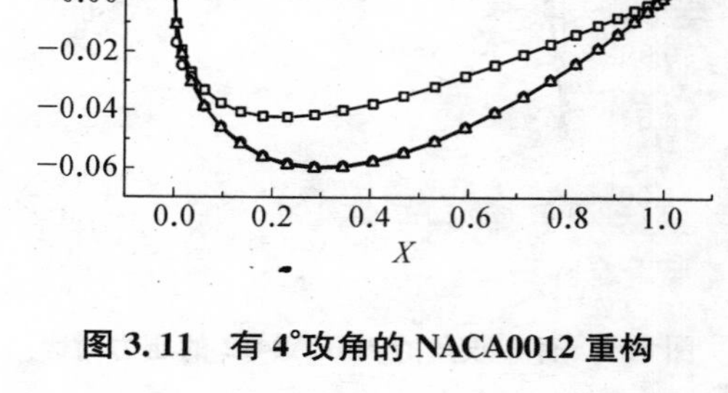

图 3.11　有 4°攻角的 NACA0012 重构

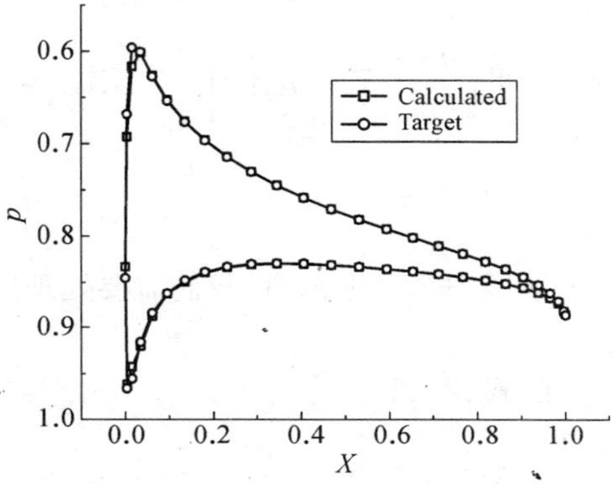

图 3.12　有 4°攻角的 NACA0012 压力对比

3.3　非定常反命题的变分泛函推导

3.3.1　从 Euler 方程反推变分原理

由前一章的推导过程可知，非定常可压缩势流的变分泛函的主要部分是由连续性方程反推出来的，即

$$\iiint\left[\frac{\partial\rho}{\partial t}+\nabla\cdot(\rho\vec{\Lambda})\right]\delta\phi\,\mathrm{d}x\mathrm{d}y\mathrm{d}t$$

$$=\frac{1}{\gamma}\delta\iiint\left\{1-\frac{1}{m}\left[\frac{\partial\phi}{\partial t}+\frac{(\nabla\phi)^2}{2}\right]\right\}^{\gamma m}\mathrm{d}x\mathrm{d}y\mathrm{d}t+$$

$$\iiint\left[\frac{\partial(\delta\phi\rho)}{\partial t}+\nabla\cdot(\rho\vec{\Lambda}\delta\phi)\right]\mathrm{d}x\mathrm{d}y\mathrm{d}t$$

也就有泛函的主要部分为：

$$J(\phi)=\frac{1}{\gamma}\iiint_{\Omega}\left\{1-\frac{1}{m}\left[\frac{\partial\phi}{\partial t}+\frac{(\nabla\phi)^2}{2}\right]\right\}^{\gamma m}\mathrm{d}x\mathrm{d}y\mathrm{d}t$$

$$=\frac{1}{\gamma}\iiint_{\Omega}p\,\mathrm{d}x\mathrm{d}y\mathrm{d}t \tag{3.29}$$

当我们选(3. 29)式作泛函时，取变分后就应为：

$$\delta J(\phi)=\iiint_{\Omega}\left[\frac{\partial\rho}{\partial t}+\nabla\cdot(\rho\vec{\Lambda})\right]\delta\phi\,\mathrm{d}x\mathrm{d}y\mathrm{d}t-\delta I_b \tag{3.30}$$

3. 3. 2 边界条件处理

由上述的计算区域图 3. 2 上知道，我们需要处理以下边界

(1) 远场边界 A_1

这与前面一章的式(2. 12)完全类似，为

$$\iint_{A_1}(\rho\Lambda_x)_{\mathrm{pr}}\phi\,\mathrm{d}s\mathrm{d}t\ \left(=\iint_{A_1}(q_n)_{\mathrm{pr}}\phi\,\mathrm{d}s\mathrm{d}t\right) \tag{3.31}$$

其中 $\rho\Lambda_n=q_n=\bar{\rho}(K_{\mathrm{I}}-K_{\mathrm{II}})/(\gamma-1)$，$K_{\mathrm{I}}$、$K_{\mathrm{II}}$ 为 Riemman 不变量，其值由∞处及前一时间层内点值求出，$\bar{\rho}$ 为上一次近似解。

(2) 翼型表面边界 A_2

由于该边界在做反问题计算时是变化的，故应使用边界上的变域变分公式

$$\iint_{A_s}\left\{G'_{n'}\delta\phi+\left[F'-G_{n'}\frac{\partial\phi}{\partial n'}\right]\delta e'_{n'}-G'_{n'}\frac{\partial\phi}{\partial\tau'}\delta e'_{\tau'}\right\}\mathrm{d}s\mathrm{d}t \tag{3.32}$$

其中

$F'=\dfrac{p}{\gamma}=\dfrac{1}{\gamma}\left\{1-\dfrac{1}{m}\left[\dfrac{\partial\phi}{\partial t}+\dfrac{(\nabla\phi)^2}{2}\right]\right\}^{\gamma m}$、$\vec{G'}=\dfrac{\partial F'}{\partial\phi_x}\vec{i}+\dfrac{\partial F'}{\partial\phi_y}\vec{j}+\dfrac{\partial F'}{\partial\phi_t}\vec{i_t}=-\rho(\vec{\Lambda}+\vec{i_t})$、$G'_{n'}=\vec{G'}\cdot\vec{n'}$，其中带上标“$'$”的都是定义在 R^3 空间中的量。$\vec{i_t}$ 为时间方向的方向向量。

我们设在 R^3 空间中的某一曲面 S 的方程为

$$f(x,\ y,\ t)=0 \tag{3.33}$$

显然，它在 R^2 空间中代表一个运动着的曲线(即一非定常曲线)。在

此曲面上,必有

$$\mathrm{d}f = \nabla' f \cdot \overrightarrow{\mathrm{d}s'} = 0 \tag{3.34}$$

其中已令

$$\begin{cases} \nabla' f = \dfrac{\partial f}{\partial x}\vec{i} + \dfrac{\partial f}{\partial y}\vec{j} + \dfrac{\partial f}{\partial t}\vec{i_t} \\ \overrightarrow{\mathrm{d}s'} = \mathrm{d}x\,\vec{i} + \mathrm{d}y\,\vec{j} + \mathrm{d}t\,\vec{i_t} \end{cases}$$

(3.34)式即变成

$$|\nabla f|\,\mathrm{d}s_n + \frac{\partial f}{\partial t}\mathrm{d}t = 0$$

注意到 $\mathrm{d}s_n/\mathrm{d}t = U$ 正好是此曲面在 R^2 中的局部无量纲推进速度 U_2,故得

$$U = \frac{\mathrm{d}s_n}{\mathrm{d}t} = -\frac{\dfrac{\partial f}{\partial t}}{|\nabla f|} \tag{3.35}$$

而且还有

$$\left(\frac{|\nabla' f|}{|\nabla f|}\right)^2 = \frac{|\nabla f|^2 + \left(\dfrac{\partial f}{\partial t}\right)^2}{|\nabla f|^2} = 1 + U^2 \tag{3.36}$$

另由 $\overrightarrow{n'} // \nabla' f$、$\vec{n} // \nabla f$,我们可得:

$$\begin{aligned} G'_{n'} &= \overrightarrow{G'} \cdot \overrightarrow{n'} = \overrightarrow{G'} \cdot \frac{\nabla' f}{|\nabla' f|} \\ &= \frac{-\rho(\vec{\Lambda} + \vec{i_t}) \cdot \left(\nabla f + \dfrac{\partial f}{\partial t}\vec{i_t}\right)}{|\nabla' f|} = -\frac{\rho\left(\vec{\Lambda} \cdot \nabla f + \dfrac{\partial f}{\partial t}\right)}{|\nabla' f|} \\ &= \frac{-\rho(\Lambda_n\,|\nabla f| - U\,|\nabla f|)}{|\nabla' f|} = \frac{\rho(U - \Lambda_n)\,|\nabla f|}{|\nabla' f|} \end{aligned}$$

$$= \frac{\rho(U - \Lambda_n)}{\sqrt{1+U^2}} \tag{3.37}$$

同理,可写出

$$\frac{\partial \phi}{\partial n'} = \nabla' \phi \cdot \overrightarrow{n'} = \frac{\nabla' \phi \cdot \left(\nabla f + \dfrac{\partial f}{\partial t} \overrightarrow{i_t} \right)}{|\nabla' f|}$$

$$= \frac{\left(\Lambda_n |\nabla f| - U \dfrac{\partial \phi}{\partial t} |\nabla f| \right)}{|\nabla' f|}$$

$$= \frac{\left(\Lambda_n - U \dfrac{\partial \phi}{\partial t} \right)}{\sqrt{1+U^2}} \tag{3.38}$$

我们还有 $\dfrac{\partial \phi}{\partial \tau'} = \Lambda_\tau$。

翼型面改变时 $\delta \overrightarrow{e'}$ 是改变的,这给计算带来了麻烦,因为我们的计算是在三维(x, y, t)空间下进行的,$\delta \overrightarrow{e'}$ 的变分最终必定要反映到静止翼型坐标(x_0, y_0)的变分上去。在非定常情况下,我们设计翼型时,为方便计,应当限制翼型变分都在同一个方向上进行,例如限制变分在静止的 y_0 方向上进行,而不是瞬时坐标,并把 x_0 作自变量,不予变分。而且,作为刚体,翼型在振荡过程中是不应该随时间变形的。

参考图 3.4 和图 3.5,设翼型的坐标变分在 $\overrightarrow{j_0}$ 方向上进行,翼型表面法向 $\vec{n}$ 如图所示。翼型振荡时这两个向量是跟随翼型一起振荡的。当静止翼型上变分量为 δy_0 时,转动一个角度后其变分量应该仍然是 δy_0,方向是转动后的 $\overrightarrow{j_0}$ 的方向。由此我们可以写成

$$\delta \overrightarrow{e'} = \delta y_0 \overrightarrow{j_0} \tag{3.39}$$

要注意的是,这里的 $\overrightarrow{j_0}$ 的方向是与时间有关系的。

由 $\vec{n'}//\nabla' f$ 、$\vec{n}//\nabla f$，我们可得：

$$\vec{n'} = \frac{\nabla' f}{|\nabla' f|} \tag{3.40}$$

则有

$$\delta \vec{e'} \cdot \vec{n'} = \delta y_0 \vec{j_0} \cdot \frac{\nabla' f}{|\nabla' f|} = \delta y_0 \vec{j_0} \cdot \frac{\vec{n}\ |\nabla f| + \frac{\partial f}{\partial t}\vec{i_t}}{|\nabla' f|} \tag{3.41}$$

由于翼型振荡过程中 $\vec{j_0}$ 始终与 $\vec{i_t}$ 方向垂直，所以我们可以得到

$$\delta \vec{e'} \cdot \vec{n'} = \delta y_0 \vec{j_0} \cdot \frac{\nabla' f}{|\nabla' f|} = \delta y_0 \vec{j_0} \cdot \vec{n} \frac{|\nabla f|}{|\nabla' f|} \tag{3.42}$$

下面我们来求 f 的表达式。由式(3.6)和(3.7)我们得到(实际上是从 (x, y) 向 (x_0, y_0) 反向转同样的 θ 角度而得)

$$x_0 = x_1 + (x - x_1)\cos\theta + (y - y_1)\sin\theta \tag{3.43}$$

$$y_0 = y_1 - (x - x_1)\sin\theta + (y - y_1)\cos\theta \tag{3.44}$$

另外，我们还有翼型的方程 $y_0 = y_0(x_0)$，经过转化我们得到了 f 的方程

$$f = y_0 - y_0(x_0) = 0 \tag{3.45}$$

从而我们有偏导数

$$\frac{\partial f}{\partial x} = \frac{\partial y_0}{\partial x} - \frac{\mathrm{d}y_0}{\mathrm{d}x_0}\frac{\partial x_0}{\partial x} = -\sin\theta - \frac{\mathrm{d}y_0}{\mathrm{d}x_0}\cos\theta \tag{3.46}$$

$$\frac{\partial f}{\partial y} = \frac{\partial y_0}{\partial y} - \frac{\mathrm{d}y_0}{\mathrm{d}x_0}\frac{\partial x_0}{\partial y} = \cos\theta - \frac{\mathrm{d}y_0}{\mathrm{d}x_0}\sin\theta \tag{3.47}$$

$$\frac{\partial f}{\partial t} = \frac{\partial y_0}{\partial t} - \frac{\mathrm{d}y_0}{\mathrm{d}x_0}\frac{\partial x_0}{\partial t}$$

$$= -(x-x_1)\cos\theta\frac{\mathrm{d}\theta}{\mathrm{d}t} - (y-y_1)\sin\theta\frac{\mathrm{d}\theta}{\mathrm{d}t} -$$
$$\frac{\mathrm{d}y_0}{\mathrm{d}x_0}[-(x-x_1)\sin\theta + (y-y_1)\cos\theta]\frac{\mathrm{d}\theta}{\mathrm{d}t}$$
$$= \left[-(x_0-x_1) - \frac{\mathrm{d}y_0}{\mathrm{d}x_0}(y_0-y_1)\right]\frac{\mathrm{d}\theta}{\mathrm{d}t} \tag{3.48}$$

则有

$$\delta\overrightarrow{e'}\cdot\overrightarrow{n'} = \delta y_0\,\overrightarrow{j_0}\cdot\vec{n}\frac{\sqrt{\left(\frac{\partial f}{\partial x}\right)^2 + \left(\frac{\partial f}{\partial y}\right)^2}}{\sqrt{\left(\frac{\partial f}{\partial x}\right)^2 + \left(\frac{\partial f}{\partial y}\right)^2 + \left(\frac{\partial f}{\partial t}\right)^2}} \tag{3.49}$$

其中

$$\left(\frac{\partial f}{\partial x}\right)^2 + \left(\frac{\partial f}{\partial y}\right)^2 = \left(-\sin\theta - \frac{\mathrm{d}y_0}{\mathrm{d}x_0}\cos\theta\right)^2 + \left(\cos\theta - \frac{\mathrm{d}y_0}{\mathrm{d}x_0}\sin\theta\right)^2$$
$$= 1 + \left(\frac{\mathrm{d}y_0}{\mathrm{d}x_0}\right)^2 \tag{3.50}$$

$$\left(\frac{\partial f}{\partial x}\right)^2 + \left(\frac{\partial f}{\partial y}\right)^2 + \left(\frac{\partial f}{\partial t}\right)^2$$
$$= 1 + \left(\frac{\mathrm{d}y_0}{\mathrm{d}x_0}\right)^2 + \left[(x_0-x_1) + \frac{\mathrm{d}y_0}{\mathrm{d}x_0}(y_0-y_1)\right]^2\left(\frac{\mathrm{d}\theta}{\mathrm{d}t}\right)^2 \tag{3.51}$$

由此我们就能得到

$$\delta\overrightarrow{e'}\cdot\overrightarrow{n'} =$$
$$\delta y_0\,\overrightarrow{j_0}\cdot\vec{n}\frac{\sqrt{1 + \left(\frac{\mathrm{d}y_0}{\mathrm{d}x_0}\right)^2}}{\sqrt{1 + \left(\frac{\mathrm{d}y_0}{\mathrm{d}x_0}\right)^2 + \left[(x_0-x_1) + \frac{\mathrm{d}y_0}{\mathrm{d}x_0}(y_0-y_1)\right]^2\left(\frac{\mathrm{d}\theta}{\mathrm{d}t}\right)^2}} \tag{3.52}$$

根据边界矢量关系式(3.18)也即有

$$\delta\overrightarrow{e'}\cdot\overrightarrow{n'}\mathrm{d}s=-\delta y_0\mathrm{d}x_0\frac{\sqrt{1+\left(\frac{\mathrm{d}y_0}{\mathrm{d}x_0}\right)^2}}{\sqrt{1+\left(\frac{\mathrm{d}y_0}{\mathrm{d}x_0}\right)^2+\left[(x_0-x_1)+\frac{\mathrm{d}y_0}{\mathrm{d}x_0}(y_0-y_1)\right]^2\left(\frac{\mathrm{d}\theta}{\mathrm{d}t}\right)^2}} \tag{3.53}$$

如果记

$$E(x_0,y_0,t)=\frac{\sqrt{1+\left(\frac{\mathrm{d}y_0}{\mathrm{d}x_0}\right)^2}}{\sqrt{1+\left(\frac{\mathrm{d}y_0}{\mathrm{d}x_0}\right)^2+\left[(x_0-x_1)+\frac{\mathrm{d}y_0}{\mathrm{d}x_0}(y_0-y_1)\right]^2\left(\frac{\mathrm{d}\theta}{\mathrm{d}t}\right)^2}},$$

则上面的变域变分公式就变成

$$\iint_{A_s}\left\{G'_{n'}\delta\phi-\left[G_{n'}\frac{\partial\phi}{\partial n'}\right]\delta e'_{n'}-G'_{n'}\frac{\partial\phi}{\partial\tau'}\delta e'_{\tau'}\right\}\mathrm{d}s\mathrm{d}t-\frac{1}{\gamma}\iint_{A_2}pE\delta y_0\mathrm{d}x_0\mathrm{d}t \tag{3.54}$$

由于 E 是 $\frac{\mathrm{d}y_0}{\mathrm{d}x_0}$ 和 y_0 的函数,我们在添加新项时需要限制其变分才行。根据式(3.11)的泛函内的压力分布,可以不难看出,要满足边界条件,需要在 J 中加入一项:

$$\frac{1}{\gamma}\int_{A_2}\left\{y_0\int_{t_{n-1}}^{t_n}[\mathring{p}+\delta(t-t_i)(p_{t_i}-\mathring{p})]\mathring{E}\mathrm{d}t\right\}\mathrm{d}x_0 \tag{3.55}$$

就能使其满足给定的时间平均压力分布。

证明:当 $t\neq t_i$ 时,式 3.54 和添加项就变成(其中 A_2' 为除 t_i 外的区域)

$$\iint_{A_s}\left\{G'_{n'}\delta\phi-\left[G_{n'}\frac{\partial\phi}{\partial n'}\right]\delta e'_{n'}-G'_{n'}\frac{\partial\phi}{\partial\tau'}\delta e'_{\tau'}\right\}\mathrm{d}s\mathrm{d}t-$$

$$\frac{1}{\gamma}\iint_{A_2'}pE\delta y_0\mathrm{d}x_0\mathrm{d}t+\frac{1}{\gamma}\iint_{A_2'}p_{\mathrm{pr}}\,\mathring{\dot{E}}\,\delta y_0\mathrm{d}x_0\mathrm{d}t$$

$$=\iint_{A_s}\left\{G'_{n'}\delta\phi-\left[G_{n'}\frac{\partial\phi}{\partial n'}\right]\delta e'_{n'}-G'_{n'}\frac{\partial\phi}{\partial\tau'}\delta e'_{\tau'}\right\}\mathrm{d}s\mathrm{d}t-$$

$$\frac{1}{\gamma}\iint_{A_2'}[pE-\mathring{p}\,\mathring{\dot{E}}]\delta y_0\mathrm{d}x_0\mathrm{d}t$$

由驻值条件立即可得自然边界条件

$$p=\mathring{p}。\tag{3.56}$$

当 $t=t_i$ 时，式(3.54)变成

$$\iint_{A_s}\left\{G'_{n'}\delta\phi-\left[G_{n'}\frac{\partial\phi}{\partial n'}\right]\delta e'_{n'}-G'_{n'}\frac{\partial\phi}{\partial\tau'}\delta e'_{\tau'}\right\}\mathrm{d}s\mathrm{d}t-$$

$$\frac{1}{\gamma}\iint_{A_2}pE\delta y_0\mathrm{d}x_0\mathrm{d}t+\frac{1}{\gamma}\iint_{A_2}p_{\mathrm{pr}}\,\mathring{\dot{E}}\,\delta y_0\mathrm{d}x_0\mathrm{d}t$$

$$=\iint_{A_s}\left\{G'_{n'}\delta\phi-\left[G_{n'}\frac{\partial\phi}{\partial n'}\right]\delta e'_{n'}-G'_{n'}\frac{\partial\phi}{\partial\tau'}\delta e'_{\tau'}\right\}\delta(t-t_i)\mathrm{d}s\mathrm{d}t-$$

$$\frac{1}{\gamma}\iint_{A_2}[pE-p_{t_i}\,\mathring{\dot{E}}]\delta y_0\delta(t-t_i)\mathrm{d}x_0\mathrm{d}t$$

也即有

$$p=p_{t_i}。\tag{3.57}$$

恰好满足了我们前面给定的某 t_i 时刻的压力分布。

由驻值条件我们还可以得到自然条件：$G'_{n'}=0$，也即

$$\rho(U-\Lambda_n)=0。\tag{3.58}$$

下面我们考虑另一种证明方法。

由 $\delta\vec{e'}=\delta y_0\vec{j_0}$，我们根据式(3.8)作分解得到

$$\delta\vec{e'}=\delta y_0\vec{j_0}=-\delta y_0\sin\theta\vec{i}+\delta y_0\cos\theta\vec{j} \tag{3.59}$$

$$\begin{aligned}\delta\vec{e'}\cdot\vec{n'}&=(-\delta y_0\sin\theta\vec{i}+\delta y_0\cos\theta\vec{j})\cdot\vec{n'}\\&=(-\delta y_0\sin\theta\vec{i}+\delta y_0\cos\theta\vec{j})\cdot\frac{\nabla' f}{|\nabla' f|}\\&=\delta y_0\frac{-\sin\theta\dfrac{\partial f}{\partial x}+\cos\theta\dfrac{\partial f}{\partial y}}{|\nabla' f|}\end{aligned} \tag{3.60}$$

把上面的几个偏导数(式 3.46、3.47、3.48)代入得到

$$-\sin\theta\frac{\partial f}{\partial x}+\cos\theta\frac{\partial f}{\partial y}=-\sin\theta\left(-\sin\theta-\frac{\mathrm{d}y_0}{\mathrm{d}x_0}\cos\theta\right)+\cos\theta\left(\cos\theta-\frac{\mathrm{d}y_0}{\mathrm{d}x_0}\sin\theta\right)=1 \tag{3.61}$$

$$\delta\vec{e'}\cdot\vec{n'}=\delta y_0\frac{1}{\sqrt{1+\left(\dfrac{\mathrm{d}y_0}{\mathrm{d}x_0}\right)^2+\left[(x_0-x_1)+\dfrac{\mathrm{d}y_0}{\mathrm{d}x_0}(y_0-y_1)\right]^2\left(\dfrac{\mathrm{d}\theta}{\mathrm{d}t}\right)^2}} \tag{3.62}$$

$$\delta\vec{e'}\cdot\vec{n'}\mathrm{d}s=\delta y_0\frac{1}{\sqrt{1+\left(\dfrac{\mathrm{d}y_0}{\mathrm{d}x_0}\right)^2+\left[(x_0-x_1)+\dfrac{\mathrm{d}y_0}{\mathrm{d}x_0}(y_0-y_1)\right]^2\left(\dfrac{\mathrm{d}\theta}{\mathrm{d}t}\right)^2}}\mathrm{d}s \tag{3.63}$$

根据转动不改变弧长的特点，我们可以把 $\mathrm{d}s$ 写成(参考图 3.6，注意

方向）

$$ds = -\sqrt{1 + \left(\frac{dy_0}{dx_0}\right)^2} dx_0 \tag{3.64}$$

从而有

$$\delta \overrightarrow{e'} \cdot \overrightarrow{n'} ds = -\delta y_0 E dx_0 \tag{3.65}$$

至此，我们可以看出这种方式推导得出的结果和前面的方式得到的结果完全一样，说明我们的推导是正确的。

(3) 尾涡边界 A_f

同上，应采用变域变分公式

$$\iint_{A_s} \left\{ G'_{n'} \delta\phi + \left[F' - G'_{n'} \frac{\partial \phi}{\partial n} \right] \delta e'_{n'} - G'_{n'} \frac{\partial \phi}{\partial \tau'} \delta e'_{\tau'} \right\} ds dt \tag{3.66}$$

由于翼型振荡过程中每个时刻的尾涡面（或尾流线）的形状都是不一样的，所以这个形状与瞬时坐标有关系。所以我们这里限制变分的方向时只能限制变分在瞬时坐标 y 向进行。根据上面的推导，我们不难得出变域变分公式变成

$$\iint_{A_f} \left\{ G'_{n'} \delta\phi ds - \left[\frac{p}{\gamma} - G'_{n'} \frac{\partial \phi}{\partial n'} \right] \frac{1}{\sqrt{1+U^2}} \delta y dx - G'_{n'} \frac{\partial \phi}{\partial \tau'} \delta e'_{\tau'} ds \right\} dt$$

对尾涡面的上下边界：

$$\iint_{A_f} \left\{ (G'_{n'} \delta\phi)_{上} ds - \left[\left(\frac{p}{\gamma} - G'_{n'} \frac{\partial \phi}{\partial n'} \right) \frac{1}{\sqrt{1+U^2}} \delta y dx \right]_{上} - \left(G'_{n'} \frac{\partial \phi}{\partial \tau'} \delta e'_{\tau'} \right)_{上} ds - (G'_{n'} \delta\phi)_{下} ds + \left[\left(\frac{p}{\gamma} - G'_{n'} \frac{\partial \phi}{\partial n'} \right) \frac{1}{\sqrt{1+U^2}} \delta y dx \right]_{下} + \left(G'_{n'} \frac{\partial \phi}{\partial \tau'} \delta e'_{\tau'} \right)_{下} ds \right\} dt \tag{3.67}$$

由于尾涡上下的 $\delta\phi_{上} \neq \delta\phi_{下}$，则可得出 $(G'_{n'})_{上} = (G'_{n'})_{下} = 0$，也即有

$$\Lambda_n = U_f。\tag{3.68}$$

由于 $y_{上} = y_{下}$，则 $\delta y_{上} = \delta y_{下}$，即有

$$\left[\left(\frac{p}{\gamma} - G'_{n'}\frac{\partial\phi}{\partial n'}\right)\delta y\right]_{上} = \left[\left(\frac{p}{\gamma} - G'_{n'}\frac{\partial\phi}{\partial n'}\right)\delta y\right]_{下}$$

也即

$$\frac{p_{上} - p_{下}}{\gamma} - \left(G'_{n'}\frac{\partial\phi}{\partial n'}\right)_{上} + \left(G'_{n'}\frac{\partial\phi}{\partial n'}\right)_{下} = 0 \tag{3.69}$$

由 $(G'_{n'})_{上} = (G'_{n'})_{下} = 0$，就可得到

$$p_{上} = p_{下}。\tag{3.70}$$

(4) 激波边界 A_s

由于它的位置是未知的，同样需要应用变域变分公式。

我们知道，激波是法向间断面，而且激波面也是与时间有关系的，所以在这里需要限制变分只在 x 向进行，当然，这也不会失去普遍性。即有：$\delta\vec{e}' = \delta x\vec{i}$，由前面的推导过程有：

$$\delta\vec{e}' \cdot \vec{n}'\mathrm{d}s = \delta x\frac{1}{\sqrt{1+U_s^2}}\mathrm{d}y \tag{3.71}$$

对激波的"+"、"−"两侧有：

$$\iint_{A_s}\Bigg\{(G'_{n'}\delta\phi)_+\,\mathrm{d}s + \left[\left(\frac{p}{\gamma} - G'_{n'}\frac{\partial\phi}{\partial n'}\right)\frac{1}{\sqrt{1+U_s^2}}\delta x\mathrm{d}y\right]_+ - \left(G'_{n'}\frac{\partial\phi}{\partial\tau'}\delta e'_{\tau'}\right)_+\mathrm{d}s - (G'_{n'}\delta\phi)_-\,\mathrm{d}s - \left[\left(\frac{p}{\gamma} - G'_{n'}\frac{\partial\phi}{\partial n'}\right)\frac{1}{\sqrt{1+U_s^2}}\delta x\mathrm{d}y\right]_- + \left(G'_{n'}\frac{\partial\phi}{\partial\tau'}\delta e'_{\tau'}\right)_-\mathrm{d}s\Bigg\}\mathrm{d}t \tag{3.72}$$

由于在激波的两侧 ϕ 值相等，即 $\phi_+ = \phi_-$，即有函数 $\psi(x, y, t) = \phi_+ - \phi_- = 0$，同样假设其为 R^3 空间上的超曲面，进行第二章中激波边界的类似推导，有

$$U_s = -\frac{\dfrac{\partial \psi}{\partial t}}{|\nabla \psi|} \tag{3.73}$$

又可得

$$-\frac{\partial \psi}{\partial t} = U_s(\nabla \phi_+ - \nabla \phi_-) = U_s(\vec{\Lambda}_+ - \vec{\Lambda}_-) \tag{3.74}$$

由 $\delta\phi_+ = \delta\phi_-$，立刻可以得出

$$(G'_{n'})_+ = (G'_{n'})_- \tag{3.75}$$

如果引入 $[|\,X\,|] = X_+ - X_-$，并把 $G'_{n'}$ 的形式代入，则有

$$[|\,\rho(\Lambda_n - U_s)\,|] = 0 \tag{3.76}$$

由 $\delta x_+ = \delta x_-$ 也可立刻导出

$$[|\,p\,|]/\gamma + \frac{\rho(\Lambda_n - U_s)}{1 + U_s^2}\left[\left|\,\Lambda_n - U_s\frac{\partial \phi}{\partial t}\right|\right] = 0$$

进一步可以写成

$$[|\,p\,|]/\gamma + \rho(\Lambda_n - U_s)[|\,\Lambda_n\,|] = 0 \tag{3.77}$$

由 $\delta\tau'_+ = \delta\tau'_-$ 可以导出

$$[|\,\Lambda_\tau\,|] = 0 \tag{3.78}$$

把式(3.73)改写为

$$\left[\left|\frac{\partial \phi}{\partial t}\right|\right] = -U_s[|\,\Lambda_n\,|]$$

把它带入能量方程后得到

$$[|H|]=(\gamma-1)U_s[|\Lambda_n|] \tag{3.79}$$

其中 H 表示总焓。我们可以看出，公式(3.76)、(3.77)、(3.78)、(3.79)组合起来恰好就是非定常势流的 Rankine-Hugoniot 激波关系式(3.13)。

值得一提的是，严格来说，上述势流控制方程是与含有激波的跨声速流动不相容的，所以实际上在出现激波后上述方程已经不再适用了，需要用赝势函数控制方程来求解。

(5) 时间边界 t_n 和 t_{n-1}

由前面的反推过程，我们得到在时间方向上只有下面一项

$$-\iiint_{\Omega}\frac{\partial(\rho\delta\phi)}{\partial t}\mathrm{d}x\mathrm{d}y\mathrm{d}t$$
$$=-\iint_{t_n}(\rho\delta\phi)_{t_n}\mathrm{d}x\mathrm{d}y+\iint_{t_{n-1}}(\rho\delta\phi)_{t_{n-1}}\mathrm{d}x\mathrm{d}y \tag{3.80}$$

当 $t=t_n$ 时，上式为

$$-\iint_{t_n}(\rho\delta\phi)\mathrm{d}x\mathrm{d}y \tag{3.81}$$

可以看出，当在原泛函中添加一项限制变分，即限制密度 ρ 的变分，以上标“o”标志

$$\iint_{t_n}\overset{\circ}{\rho}\phi\mathrm{d}x\mathrm{d}y \tag{3.82}$$

就可以满足时间 t_n 边界上的条件要求，并且保证了双曲型方程在 t_n 时间上不能给终值条件的要求。

当 $t=t_{n-1}$ 时：剩余项为

$$\iint_{t_{n-1}}(\rho\delta\phi)\mathrm{d}x\mathrm{d}y \tag{3.83}$$

引进拉氏乘子 μ_1、μ_2 把 $\rho=\rho_{pr}$ 和 $\phi=\phi_{pr}$ 两边界条件引入

$$\iint_{t_{n-1}}\{\rho\delta\phi+\delta[\mu_1(\rho-\rho_{pr})]+\delta[\mu_2(\phi-\phi_{pr})]\}\mathrm{d}x\mathrm{d}y$$
$$=\iint_{t_{n-1}}[\rho\delta\phi+(\rho-\rho_{pr})\delta\mu_1+\mu_1\delta\rho+(\phi-\phi_{pr})\delta\mu_2+\mu_2\delta\phi]\mathrm{d}x\mathrm{d}y$$
$$=\iint_{t_{n-1}}[(\rho+\mu_2)\delta\phi+(\rho-\rho_{pr})\delta\mu_1+\mu_1\delta\rho+(\phi-\phi_{pr})\delta\mu_2]\mathrm{d}x\mathrm{d}y$$

由于各自独立变分,从而可得

$$\mu_1=0\text{ , }\mu_2=-\rho \tag{3.84}$$

由于 $\mu_1=0$,采用第二章中的类似处理,令 $\mu_1=\frac{1}{2}(\phi-\phi_{pr})$,加入到泛函中后,有

$$\frac{1}{2}(\phi-\phi_{pr})(\rho-\rho_{pr})-\rho(\phi-\phi_{pr})$$
$$=\frac{1}{2}(\phi\rho-\phi\rho_{pr}-\rho\phi_{pr}+\phi_{pr}\rho_{pr})-\rho\phi+\rho\phi_{pr}$$
$$=\frac{1}{2}\rho\phi_{pr}-\frac{1}{2}\phi\rho-\frac{1}{2}\phi\rho_{pr}+\frac{1}{2}\phi_{pr}\rho_{pr}$$

去掉常数项得

$$-\frac{1}{2}[\rho_{pr}\phi+(\phi-\phi_{pr})\rho] \tag{3.85}$$

故应在原泛函中加入的项为

$$-\frac{1}{2}\iint_{t_{n-1}}[\rho_{pr}\phi+(\phi-\phi_{pr})\rho]\mathrm{d}x\mathrm{d}y \tag{3.86}$$

这是逐时间层推进时的 t_{n-1} 时刻的边界条件。对于初始时刻 t_0 来说,式中的 ϕ_{pr} 和 ρ_{pr} 就变成初始边界条件中的 f_2 和 f_3。

(6) 非定常流动的 Kutta 条件

采用变域变分有限元方法可以把尾涡面的形状求解出来，满足了尾涡面的边界条件。对于尾缘点的非定常 Kutta 条件，我们这里采用无载荷条件。我们只知道尾涡面的起始点在翼型表面，具体位置需要通过尾缘点的无载荷条件来决定。针对这种情况，可以用牛顿迭代法来决定，迭代过程和翼型定常绕流问题的势函数差的求解过程类似(参考附录 B)。即先假定一个尾涡面的起始点，一般取在尾缘点附近，视振荡的角度定在翼型的哪一侧，然后计算尾缘点上下的压力差，根据连续两次得到的压力差可以得到尾涡面新起始点的位置。

3.4 非定常反命题的变分原理

上述振荡翼型非定常绕流 $\mathrm{I_B}$ 型反命题的解使下列泛函 J_1 取驻值：$\delta J_1 = 0$。其中 ϕ、A_2、A_s 和 A_f 各自独立变分，

$$J_1(\phi, A_2, A_s, A_f)$$
$$= \frac{1}{\gamma}\iiint\limits_{\Omega}\left\{1-(\gamma-1)\left[\frac{\partial\phi}{\partial t}+\frac{(\nabla\phi)^2}{2}\right]\right\}^{\frac{\gamma}{\gamma-1}}\mathrm{d}x\mathrm{d}y\mathrm{d}t + L \quad (3.87)$$

其中

$$L = \iint\limits_{A_1}(q_n)_{\mathrm{pr}}\phi\,\mathrm{d}s\mathrm{d}t + L_b + L_T$$

$$L_b = \frac{1}{\gamma}\int\limits_{A_2}\left\{y_0\int_{t_{n-1}}^{t_n}\left[\mathring{p}+\delta(t-t_i)(p_{t_i}-\mathring{p})\right]\mathring{E}\mathrm{d}t\right\}\mathrm{d}x_0$$

$$L_T = \iint\limits_{t_n}\mathring{\rho}\phi\,\mathrm{d}x\mathrm{d}y - \frac{1}{2}\iint\limits_{t_{n-1}}\left[\rho_{\mathrm{pr}}\phi+(\phi-\phi_{\mathrm{pr}})\rho\right]\mathrm{d}x\mathrm{d}y$$

$$E = \frac{\sqrt{1+\left(\dfrac{\mathrm{d}y_0}{\mathrm{d}x_0}\right)^2}}{\sqrt{1+\left(\dfrac{\mathrm{d}y_0}{\mathrm{d}x_0}\right)^2+\left[(x_0-x_1)+\dfrac{\mathrm{d}y_0}{\mathrm{d}x_0}(y_0-y_1)\right]^2\left(\dfrac{\mathrm{d}\theta}{\mathrm{d}t}\right)^2}}$$

3.5 变分原理的推广

根据变分推导的系统性途径，借助于拉氏乘子，不难将上述变分原理推广得出一系列的广义变分原理及亚广义变分原理族。

3.5.1 亚广义变分原理

振荡翼型非定常绕流 $\mathrm{I_B}$ 型反命题的解使下列泛函 J_2 取驻值：$\delta J_2 = 0$，其中

$$J_2(\phi, \vec{\Lambda}, A_2, A_s, A_f) = -\iiint_{\Omega}\left[1-(\gamma-1)\left(\frac{\partial\phi}{\partial t}+\frac{\Lambda^2}{2}\right)\right]^{\frac{1}{\gamma-1}} \cdot \left[\vec{\Lambda}\cdot\nabla\phi+\frac{\gamma-1}{\gamma}\frac{\partial\phi}{\partial t}-\frac{1}{\gamma}\left(1+\frac{\gamma+1}{2}\Lambda^2\right)\right]\mathrm{d}V+L \tag{3.88}$$

L 式同式(3.87)。

3.5.2 广义变分原理

振荡翼型非定常绕流 $\mathrm{I_B}$ 型反命题的解使下列泛函 J_3 取驻值：$\delta J_3 = 0$，其中

$$J_3(\phi, \vec{\Lambda}, \rho, p, A_2, A_s, A_f) = \iiint_{\Omega}\left\{\rho\left(\frac{\Lambda^2}{2}-\frac{\partial\phi}{\partial t}-\vec{\Lambda}\cdot\nabla\phi\right)-\frac{1}{\gamma-1}\left[\frac{p}{\gamma}\left(1-\ln\frac{p}{\rho^{\gamma}}\right)-\rho\right]\right\}\mathrm{d}V+L \tag{3.89}$$

L 式同式(3.87)。

3.5.3 广义变分原理的普遍形式

应用线性组合法，可在上列诸 VP 基础上，构造出下列 GVP 的普遍形式：振荡翼型非定常绕流 $\mathrm{I_B}$ 型反命题的解使下列泛函 J_4 取驻值：$\delta J_4 = 0$

$$\begin{aligned} & J_4(\phi,\ \vec{\Lambda},\ \rho,\ p,\ A_2,\ A_f,\ A_s,\ C_1,\ C_2) \\ & = C_1 J_1 + C_2 J_2 + (1 - C_1 - C_2) J_3 \end{aligned} \tag{3.90}$$

L 式同式(3.87)。

3.6 计算步骤

(1) 初始化某些必要参数，如远场条件，初始流场，并生成初始网格；

(2) 置周期指针 IT=1；

(3) 判断是否是求反问题的周期，如是则把存储 $\dfrac{\partial J}{\partial y_{0j}}$ 的数组清零；

(4) 置当前时间层 K=1；

(5) 初始化本时间层参数，以上一层的流场作初值；

(6) 计算势函数、密度、速度等变量；

(7) 求本时间层的尾涡位置，用 $\dfrac{\partial y}{\partial t} = \dfrac{\partial J}{\partial y}$ 伪非定常法；

(8) 计算本时间层与上一周期同一层的剩余以及 Kutta 条件的满足情况；

(9) 判断剩余是否在一定范围内，如是，(看是否这周期是求反问题的周期，如是，则计算本时间层的 $\dfrac{\partial J}{\partial y_{0j}}$，加到本周期上存储 $\dfrac{\partial J}{\partial y_{0j}}$ 的数组上去。)，就看是否一个整周期结束，是则转(10)，否则置 K=K+1，转(5)。否则转(6)；

(10) 一个周期结束。如是求反问题的周期，则求一次反问题，改变翼型，生成新的网格（所有时间层上的）。否则直接转(11)；

(11) 计算本周期的剩余，在一定范围内则停止迭代，转(12)。否则置 IT=IT+1，转(3)；

(12) 打印输出。

具体计算的流程图可以看下一章的图 4.1。

3.7 小结

本章首先提出了非定常反问题的问题描述，考虑了当前计算区域的各种边界条件。为了更容易地理解非定常反问题的变分过程，我们先采用了一种新方式来推导了定常反问题的变域变分原理，并以新的变分泛函设计了一个对称和一个非对称翼型，都得到了相当满意的结果。在此基础上，我们对非定常反设计的变分泛函进行了详细地推导。在进行翼型坐标变域变分的推导过程中我们采取了两种方式，得到了完全一致的结果。本文对边界条件经过处理后得到了完整的变分原理。在此基础上还把该变分原理进行了广义化，得到了相应的广义、亚广义变分原理以及广义变分原理的普遍形式。最后还探讨了该种反问题的求解步骤。

第四章　振荡翼型正、反命题的变域变分全时空有限元解

4.1　非定常正命题求解

实际上我们在前一章中已经得到了非定常振荡翼型绕流的变分泛函。只需要去掉反问题变分泛函式中间的翼型表面的添加项，并且不让 A_2 自由变分就行，具体泛函为

$$
\begin{aligned}
&J_1(\phi,\ A_s,\ A_f)\\
&=\frac{1}{\gamma}\iiint\limits_{\Omega}\left\{1-(\gamma-1)\left[\frac{\partial\phi}{\partial t}+\frac{(\nabla\phi)^2}{2}\right]\right\}^{\frac{\gamma}{\gamma-1}}\mathrm{d}x\mathrm{d}y\mathrm{d}t+L
\end{aligned}
\tag{4.1}
$$

其中

$$
L=\iint\limits_{A_1}(q_n)_{\mathrm{pr}}\phi\,\mathrm{d}s\mathrm{d}t+L_T
$$

$$
L_T=\iint\limits_{t_n}\overset{\circ}{\rho}\phi\,\mathrm{d}x\mathrm{d}y-\frac{1}{2}\iint\limits_{t_{n-1}}[\rho_{\mathrm{pr}}\phi+(\phi-\phi_{\mathrm{pr}})\rho]\mathrm{d}x\mathrm{d}y
$$

4.1.1　势函数求解离散

同第二章，对势函数在有限单元内作如下的离散：

$$
\phi(\xi,\ \eta,\ \zeta)=N_i[\phi_i+a_i(1+\zeta)]\ (\text{遵守求和约定}) \tag{4.2}
$$

其中

$$N_1 = \frac{1-\xi}{2}\frac{1-\eta}{2}, N_2 = \frac{1+\xi}{2}\frac{1-\eta}{2},$$
$$N_3 = \frac{1+\xi}{2}\frac{1+\eta}{2}, N_4 = \frac{1-\xi}{2}\frac{1+\eta}{2}。$$

定义从空间-时间坐标 (x, y, t) 到有限单元体积坐标 (ξ, η, ζ) 之间的 Jacobi 转换矩阵如下

$$J = \begin{bmatrix} \frac{\partial x}{\partial \xi} & \frac{\partial y}{\partial \xi} & \frac{\partial t}{\partial \xi} \\ \frac{\partial x}{\partial \eta} & \frac{\partial y}{\partial \eta} & \frac{\partial t}{\partial \eta} \\ \frac{\partial x}{\partial \zeta} & \frac{\partial y}{\partial \zeta} & \frac{\partial t}{\partial \zeta} \end{bmatrix} \tag{4.3}$$

从而有单元内势函数对空间-时间的偏导数

$$\begin{bmatrix} \frac{\partial \phi}{\partial x} \\ \frac{\partial \phi}{\partial y} \\ \frac{\partial \phi}{\partial t} \end{bmatrix} = J^{-1} \begin{bmatrix} \frac{\partial \phi}{\partial \xi} \\ \frac{\partial \phi}{\partial \eta} \\ \frac{\partial \phi}{\partial \zeta} \end{bmatrix} = \begin{bmatrix} J_1^{-1} \\ J_2^{-1} \\ J_3^{-1} \end{bmatrix} \begin{bmatrix} \frac{\partial \phi}{\partial \xi} \\ \frac{\partial \phi}{\partial \eta} \\ \frac{\partial \phi}{\partial \zeta} \end{bmatrix} = \begin{bmatrix} J_1^{-1} \\ J_2^{-1} \\ J_3^{-1} \end{bmatrix} \nabla' \phi \tag{4.4}$$

其中 J^{-1} 为 Jacobi 矩阵的逆矩阵，J_1^{-1}、J_2^{-1}、J_3^{-1} 分别为由 Jacobi 矩阵的逆矩阵的第一、二、三行元素组成的行向量。$\nabla' \phi$ 为势函数对体积坐标的散度，分别为

$$\frac{\partial \phi}{\partial \xi} = \frac{\partial N_i}{\partial \xi}[\phi_i + a_i(1+\zeta)] \tag{4.5a}$$

$$\frac{\partial \phi}{\partial \eta} = \frac{\partial N_i}{\partial \eta}[\phi_i + a_i(1+\zeta)] \tag{4.5b}$$

$\frac{\partial \phi}{\partial \zeta} = N_i a_i$（遵守求和约定），$i = 1, 2, 3, 4$（下面对 j, k 同）。

(4.5c)

由此我们可得到单元内泛函对任意 a_i 的偏导数为

$$\frac{\partial J_{1e}}{\partial a_i}=-\iiint_{\Omega}\rho\left[J_3^{-1}\begin{bmatrix}\frac{\partial N_i}{\partial \xi}(1+\zeta)\\ \frac{\partial N_i}{\partial \eta}(1+\zeta)\\ N_i\end{bmatrix}+J_1^{-1}\begin{bmatrix}\frac{\partial N_j}{\partial \xi}[\phi_j+a_j(1+\zeta)]\\ \frac{\partial N_j}{\partial \eta}[\phi_j+a_j(1+\zeta)]\\ N_j a_j\end{bmatrix}\cdot\right.$$

$$J_1^{-1}\begin{bmatrix}\frac{\partial N_i}{\partial \xi}(1+\zeta)\\ \frac{\partial N_i}{\partial \eta}(1+\zeta)\\ N_i\end{bmatrix}+J_2^{-1}\begin{bmatrix}\frac{\partial N_j}{\partial \xi}[\phi_j+a_j(1+\zeta)]\\ \frac{\partial N_j}{\partial \eta}[\phi_j+a_j(1+\zeta)]\\ N_j a_j\end{bmatrix}\cdot$$

$$\left.J_2^{-1}\begin{bmatrix}\frac{\partial N_i}{\partial \xi}(1+\zeta)\\ \frac{\partial N_i}{\partial \eta}(1+\zeta)\\ N_i\end{bmatrix}\right]\cdot|J|\,\mathrm{d}\xi\mathrm{d}\eta\mathrm{d}\zeta+\iint_{A_1}(\rho\Lambda_n)[N_i(1+\zeta)]\cdot$$

$$E_{13}\,\mathrm{d}\xi\mathrm{d}\zeta+\iint_{t_n}\mathring{\rho}N_i(1+\zeta)\cdot E_{12}\,\mathrm{d}\xi\mathrm{d}\eta \tag{4.6}$$

上式中对 j 和 k 有求和约定。上列诸式中 $|J|$ 为 Jacobi 矩阵的行列式，E_{12}、E_{13} 为面积分的转换系数，具体形式见第二章中式(2.35)，式(2.36)。

为了把上式中含 a 的项线性化处理，我们把 $\frac{\partial \phi}{\partial x}$ 和 $\frac{\partial \phi}{\partial y}$ 中的 a 提出来，从而可以得到单元内写成矩阵形式的代数式：

$$\frac{\partial J_{1e}}{\partial a_i}=-\begin{bmatrix}b_{11}&b_{12}&b_{13}&b_{14}\\ b_{21}&b_{22}&b_{23}&b_{24}\\ b_{31}&b_{32}&b_{33}&b_{34}\\ b_{41}&b_{42}&b_{43}&b_{44}\end{bmatrix}\begin{bmatrix}a_1\\ a_2\\ a_3\\ a_4\end{bmatrix}+\begin{bmatrix}p_1\\ p_2\\ p_3\\ p_4\end{bmatrix} \tag{4.7}$$

其中的参数为

$$b_{ij}=\iiint_{\Omega}\rho\left[J_1^{-1}\begin{bmatrix}\frac{\partial N_i}{\partial\xi}(1+\zeta)\\ \frac{\partial N_i}{\partial\eta}(1+\zeta)\\ N_i\end{bmatrix}\cdot J_1^{-1}\begin{bmatrix}\frac{\partial N_j}{\partial\xi}(1+\zeta)\\ \frac{\partial N_j}{\partial\eta}(1+\zeta)\\ N_j\end{bmatrix}+\right.$$

$$\left.J_2^{-1}\begin{bmatrix}\frac{\partial N_i}{\partial\xi}(1+\zeta)\\ \frac{\partial N_i}{\partial\eta}(1+\zeta)\\ N_i\end{bmatrix}\cdot J_2^{-1}\begin{bmatrix}\frac{\partial N_j}{\partial\xi}(1+\zeta)\\ \frac{\partial N_j}{\partial\eta}(1+\zeta)\\ N_j\end{bmatrix}\right]\cdot|J|\,\mathrm{d}\xi\mathrm{d}\eta\mathrm{d}\zeta$$

$$p_i=-\iiint_{\Omega}\rho\left[J_3^{-1}\begin{bmatrix}\frac{\partial N_i}{\partial\xi}(1+\zeta)\\ \frac{\partial N_i}{\partial\eta}(1+\zeta)\\ N_i\end{bmatrix}+J_1^{-1}\begin{bmatrix}\frac{\partial N_k}{\partial\xi}\phi_k\\ \frac{\partial N_k}{\partial\eta}\phi_k\\ 0\end{bmatrix}\cdot J_1^{-1}\begin{bmatrix}\frac{\partial N_i}{\partial\xi}(1+\zeta)\\ \frac{\partial N_i}{\partial\eta}(1+\zeta)\\ N_i\end{bmatrix}+\right.$$

$$\left.J_2^{-1}\begin{bmatrix}\frac{\partial N_k}{\partial\xi}\phi_k\\ \frac{\partial N_k}{\partial\eta}\phi_k\\ 0\end{bmatrix}\cdot J_2^{-1}\begin{bmatrix}\frac{\partial N_i}{\partial\xi}(1+\zeta)\\ \frac{\partial N_i}{\partial\eta}(1+\zeta)\\ N_i\end{bmatrix}\right]\cdot|J|\,\mathrm{d}\xi\mathrm{d}\eta\mathrm{d}\zeta+$$

$$\iint_{A_1}(\rho\Lambda_n)[N_i(1+\zeta)]\cdot E_{13}\,\mathrm{d}\xi\mathrm{d}\zeta+\iint_{t_n}\mathring{\rho}N_i(1+\zeta)\cdot E_{12}\,\mathrm{d}\xi\mathrm{d}\eta$$

p 式中对 k 有求和约定。

合并各单元的上述代数式可以得到整个网格上的关于 a 的方程。

初始条件取初始密度和势函数。假设翼型开始时并不振荡，处于平均攻角位置。则我们可以对翼型的定常绕流进行分析，得到相应的密度和势函数。由此，我们也就得到了振荡翼型初始时刻的密度和势函数。由于我们的计算方法是隐式的，可以取较大的时间步长。同第二章中的给定 a 值的方法类似，我们这里仍然以特征面公式得到远场的速度和密度，从而可以算出当地的 a 值。具体的求解公式见第二章中的式(2.39)～式(2.42)。

4.1.2 尾涡面位置求解

尾涡面的求解实际上是非定常的变域变分的求解过程。所以，尾涡面的求解与定常的变域变分的求解类似，但离散过程有所不同。

这里的离散仍然需要用到式(4.2)、(4.3)、(4.4)、(4.5)。

在单元中泛函对尾涡面瞬时 y 坐标的偏导数为

$$\frac{\partial J_{1e}}{\partial y_i}=-\iiint_{\Omega}\rho\left[\frac{\partial J_3^{-1}}{\partial y_i}\nabla'\phi+\frac{\partial J_1^{-1}}{\partial y_i}\nabla'\phi\cdot J_1^{-1}\nabla'\phi+\frac{\partial J_2^{-1}}{\partial y_i}\nabla'\phi\cdot J_2^{-1}\nabla'\phi\right]|J|\,\mathrm{d}\xi\mathrm{d}\eta\mathrm{d}\zeta+\iiint_{\Omega}\frac{p}{\gamma}\cdot\frac{\partial|J|}{\partial y_i}\mathrm{d}\xi\mathrm{d}\eta\mathrm{d}\zeta+\iint_{t_n}\mathring{\rho}[\phi_k+a_k(1+\zeta)]\frac{\partial E_{12}}{\partial y_i}\mathrm{d}\xi\mathrm{d}\eta$$

把(4.4)、(4.5)代入后得到

$$\frac{\partial J_{1e}}{\partial y_i}=-\iiint_{\Omega}\rho\left[\frac{\partial J_3^{-1}}{\partial y_i}\begin{bmatrix}\frac{\partial N_j}{\partial\xi}[\phi_j+a_j(1+\zeta)]\\ \frac{\partial N_j}{\partial\eta}[\phi_j+a_j(1+\zeta)]\\ N_ja_j\end{bmatrix}+\frac{\partial J_1^{-1}}{\partial y_i}\begin{bmatrix}\frac{\partial N_j}{\partial\xi}[\phi_j+a_j(1+\zeta)]\\ \frac{\partial N_j}{\partial\eta}[\phi_j+a_j(1+\zeta)]\\ N_ja_j\end{bmatrix}\cdot\right.$$

$$J_1^{-1}\begin{bmatrix}\dfrac{\partial N_k}{\partial \xi}[\phi_k + a_k(1+\zeta)] \\ \dfrac{\partial N_k}{\partial \eta}[\phi_k + a_k(1+\zeta)] \\ N_k a_k\end{bmatrix} + \frac{\partial J_2^{-1}}{\partial y_i}\begin{bmatrix}\dfrac{\partial N_j}{\partial \xi}[\phi_j + a_j(1+\zeta)] \\ \dfrac{\partial N_j}{\partial \eta}[\phi_j + a_j(1+\zeta)] \\ N_j a_j\end{bmatrix} \cdot$$

$$\left. J_2^{-1}\begin{bmatrix}\dfrac{\partial N_k}{\partial \xi}[\phi_k + a_k(1+\zeta)] \\ \dfrac{\partial N_k}{\partial \eta}[\phi_k + a_k(1+\zeta)] \\ N_k a_k\end{bmatrix}\right] \cdot |J| \, \mathrm{d}\xi \mathrm{d}\eta \mathrm{d}\zeta +$$

$$\iiint_{\Omega} \frac{p}{\gamma} \cdot \frac{\partial |J|}{\partial y_i} \mathrm{d}\xi \mathrm{d}\eta \mathrm{d}\zeta + \iint_{t_n} \mathring{\rho}[\phi_k + a_k(1+\zeta)] \frac{\partial E_{12}}{\partial y_i} \mathrm{d}\xi \mathrm{d}\eta \qquad (4.8)$$

其中 $\dfrac{\partial J_1^{-1}}{\partial y_i}$、$\dfrac{\partial J_2^{-1}}{\partial y_i}$、$\dfrac{\partial J_3^{-1}}{\partial y_i}$ 的求法需要根据 J^{-1} 的推导过程得到其表达式,然后再求该表达式对 y_i 的导数。同样,$\dfrac{\partial |J|}{\partial y_i}$ 和 $\dfrac{\partial E_{12}}{\partial y_i}$ 也需要根据表达式求出。具体求法可以参考文献[120]。

得到了 $\dfrac{\partial J_{1e}}{\partial y_i}$ 的值后同样需要经过总体的合成得到区域中所有单元内的泛函对该点坐标的偏导数。进而可以通过前一章中提到的伪非定常的方法(即参考式 3.26 和 3.27)求解翼型尾流线的瞬时坐标。由于该问题的非线性比较强,这里的时间步长就要取得比较小,经过计算表明在 0.1～0.8 范围内是比较理想的。为了保证收敛,我们设置每求解势函数 6 次求解一次尾流线的位置,为的是让得到的势函数比较稳定。

4.1.3 求解流程

如图 4.1 所示。图中的 IPD 表示计算周期数。IK 表示本周期内的时间层数。IT 则表示每周期内的迭代次数。计算时总的周期数在 10 次左右就收敛得很好了。一周期内分了 8 个时间层进行计算。计算一个周期在 PIV1. 7G 上大约需要 50 min。

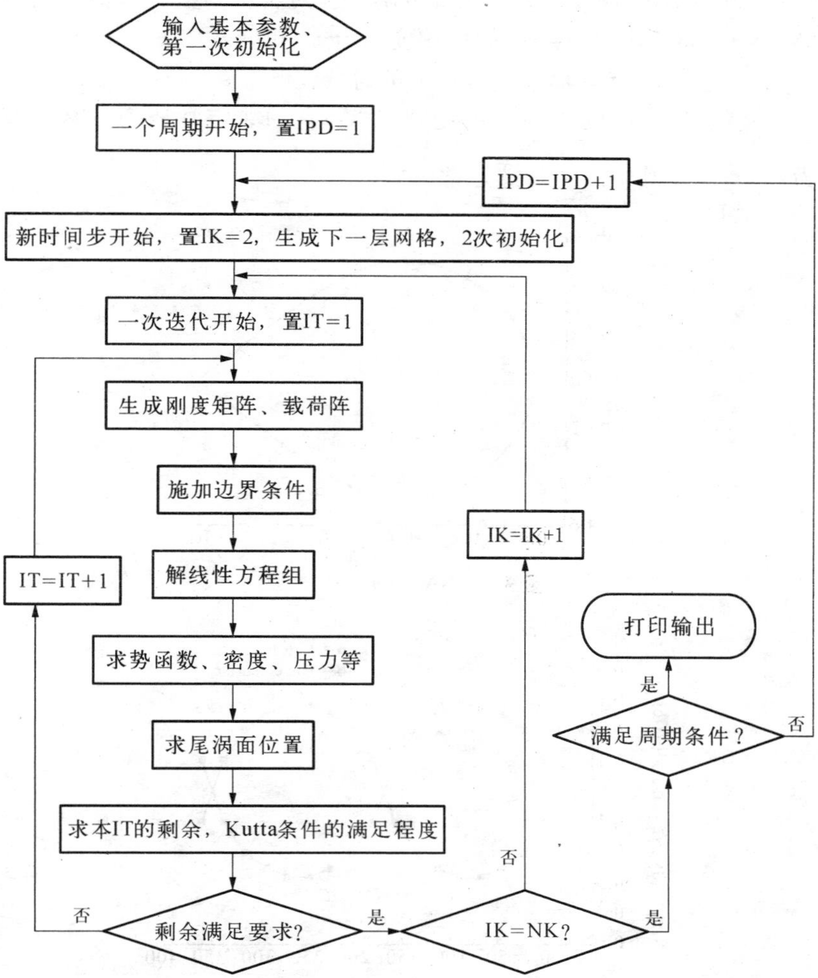

图 4.1　非定常计算流程图

4.1.4 算例

我们这里的第一个算例选自文献[114]。文中测试的是NACA64A010翼型(参见图4.2)的绕流,其实验参数为来流马赫数$Ma=0.168$。攻角$\alpha=0°$,振荡幅度$A=1°$,折合频率$k=0.49$。

(1) 振荡翼型表面非定常压力的时域波形

文献[114]中的试验结果见图4.3。它的非定常压力系数定义为:$\dfrac{p}{\frac{1}{2}\rho U_\infty^2}$。其中的$p$为非定常压力。

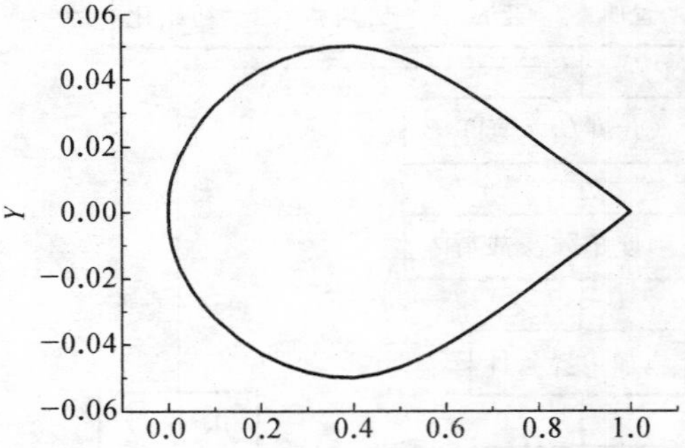

图4.2 NACA64A010翼型

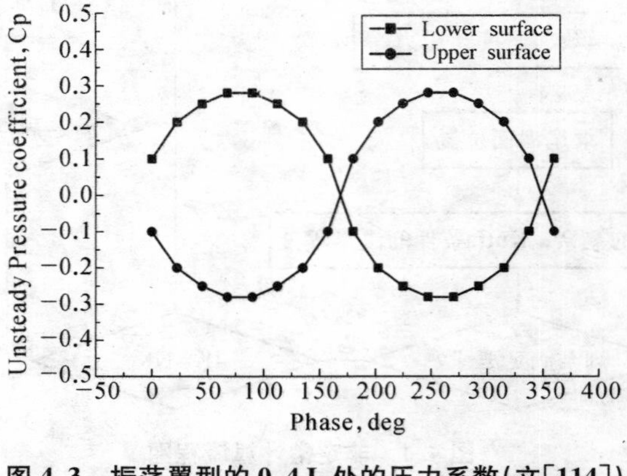

图4.3 振荡翼型的0.4L处的压力系数(文[114])

本文中的压力系数随时间的变化可从后面的整周期压力系数比较图上看出。

(2) 翼型的尾涡形状

从图 4.4 中我们看出尾涡面的收敛情况是很不错的。在图 4.5 中我们看出计算得到的周期性满足得相当好。而且从图上还可以看到尾流线的出流是与翼型的下表面相切的。与 Giesing-Maskell 流动模型吻合得相当好,也进一步验证了该模型的正确性。我们从图 4.5 中看到,翼型处于平衡位置时其尾流线并不是水平的。由于我们设置了翼型从一开始就做顺时针摆动,所以尾流线是向上的。从图 4.6 上看出,翼型振荡往返平衡位置时其尾流线上下对称,周期性满足得相当好。从图 4.7 上可知,在非平衡位置时,往返同一位置其尾流线相差很大。图 4.8 是经过平移后的尾涡线,可以看出周期性关系满足得很好。

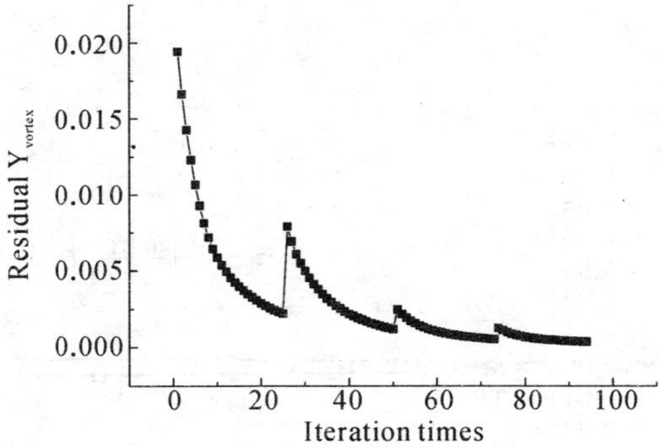

图 4.4 尾涡面的收敛过程(四个周期)

(3) 翼型表面的压力系数

从图 4.9 中可以很明显地看出,在平衡位置时翼型表面是有升力的,与静止时的情况不同。图 4.10 的半周期对比表明,上下翼型面的压力系数刚好完全交换位置。而且从图 4.9 和图 4.10 对比发现,计算的周期性满足得很好。从图 4.11 上可以看出往返同一非平衡位置

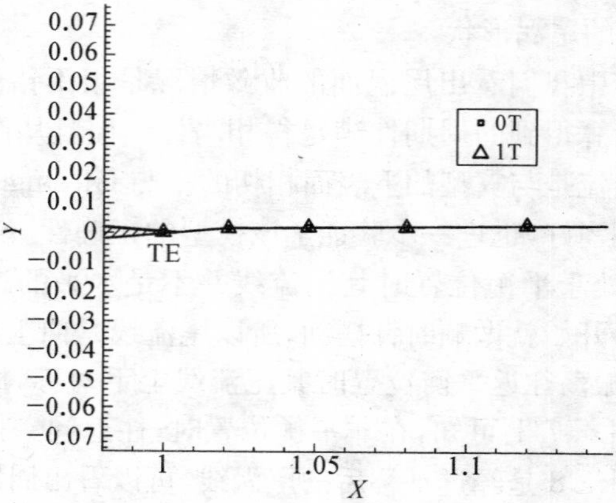

图 4.5　一个周期尾涡线对比

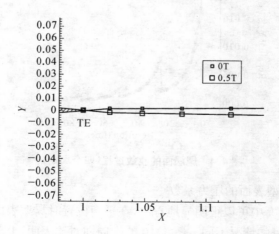

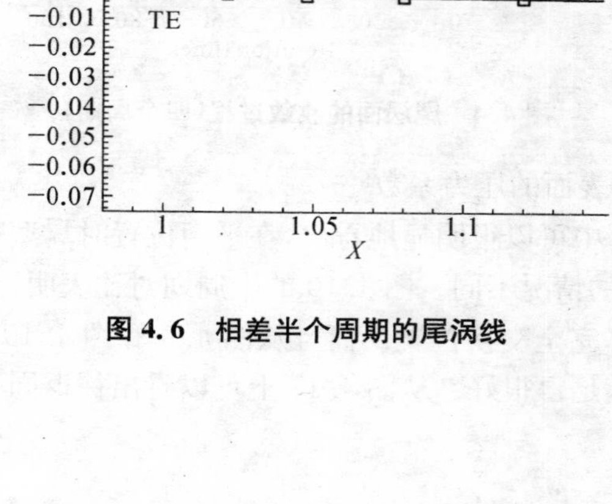

图 4.6　相差半个周期的尾涡线

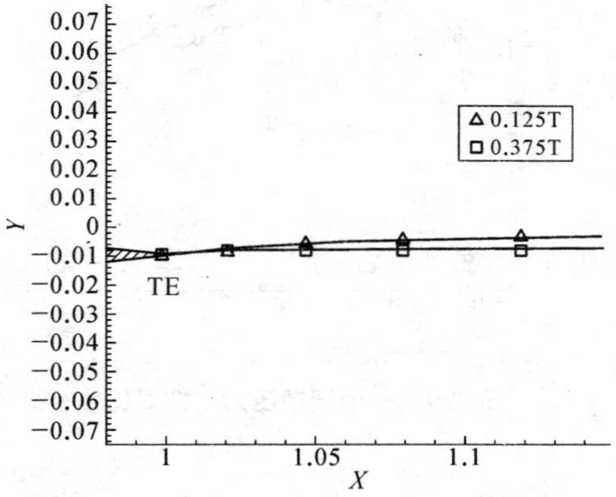

图 4.7　往返同一位置的尾涡线

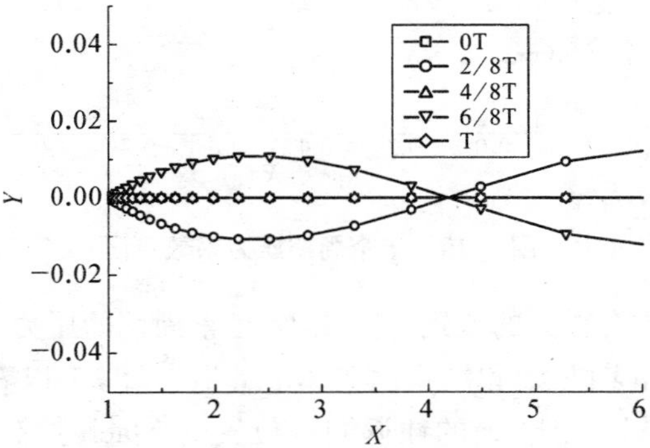

图 4.8　经过平移后几个时刻的尾涡线

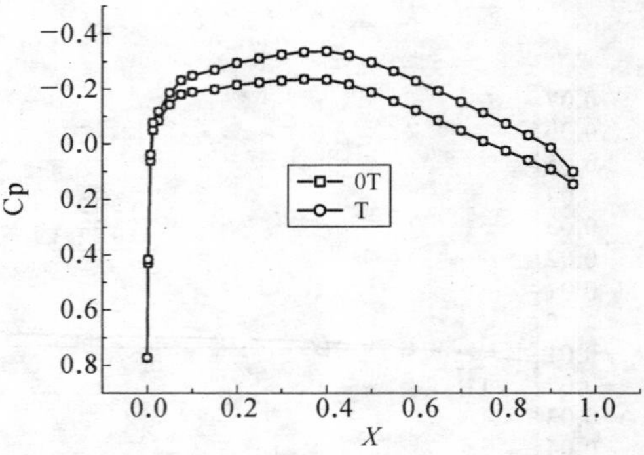

图 4.9　一个周期压力系数对比

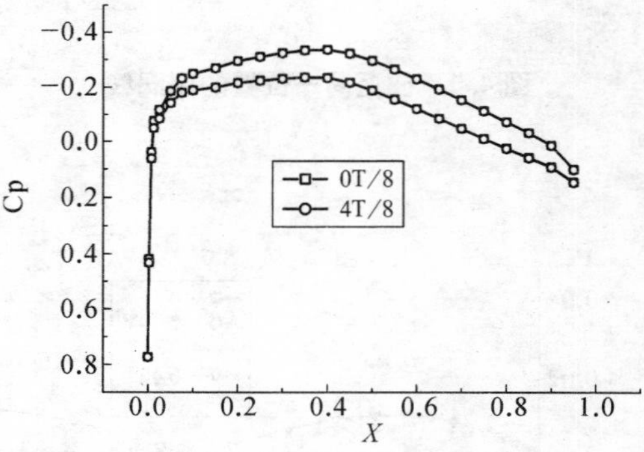

图 4.10　半个周期压力系数对比

时其压力系数有很大的差别。图 4.12 是整周期的压力系数对比。图 4.13 是一个周期开始和结束时的 Mach 数对比，可以看出它们符合得相当好。从图 4.14 上的往返同一位置的 Mach 数对比结果发现，它们是有较大差别的，与压力系数比较的情况相同。

我们的第二个算例是考虑的非对称的 NACA2412 翼型的振荡情况。流动控制参数仍然与上面的相同，即攻角为 0°，振幅为 1°，折合

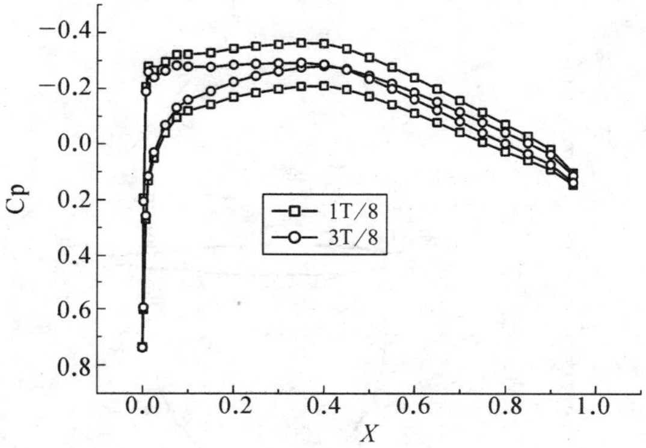

图 4.11　往返同一位置压力系数对比

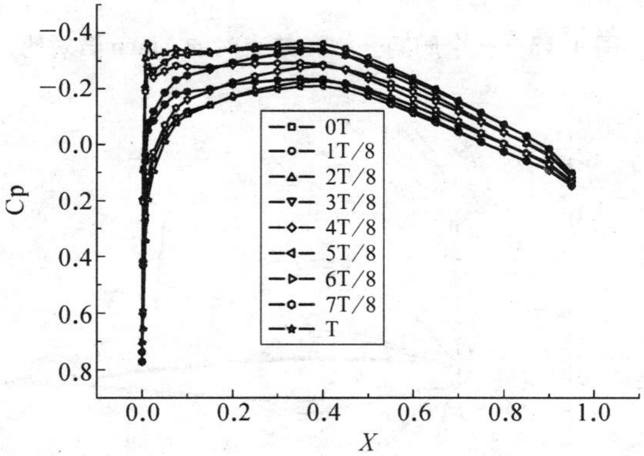

图 4.12　整周期压力系数对比

频率为 0.49，来流 Mach 数为 0.168。图 4.15 是振荡一个周期初始和终了时刻的压力系数对比，图 4.16 是振荡半个周期的压力系数对比，图 4.17 是振荡往返同一位置时的压力系数对比。由图 4.15 我们很清楚地看到我们的计算得到的周期性满足得相当好。从半个振荡周期的压力系数对比图 4.16 看出，对于非对称的振荡翼型绕流，其往

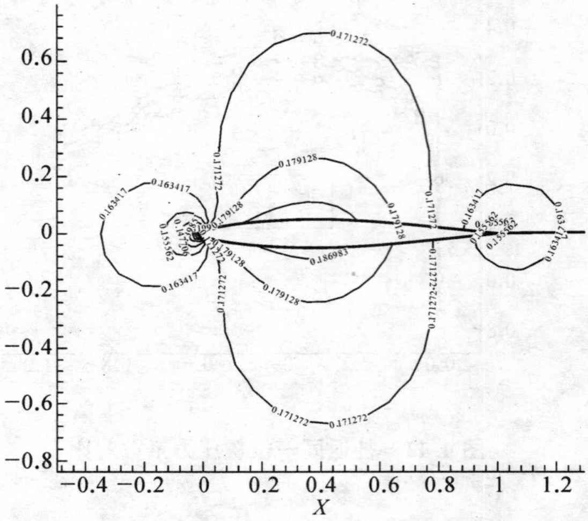

图 4.13　一个周期开始和结束时的 Mach 数对比

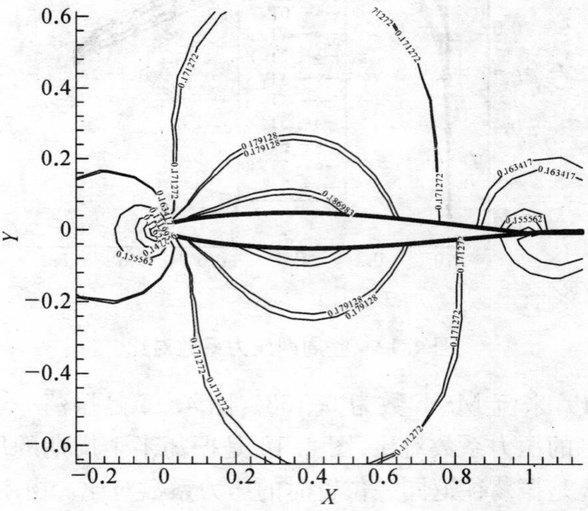

图 4.14　往返同一位置时 Mach 数对比(1T/8 和 3T/8)

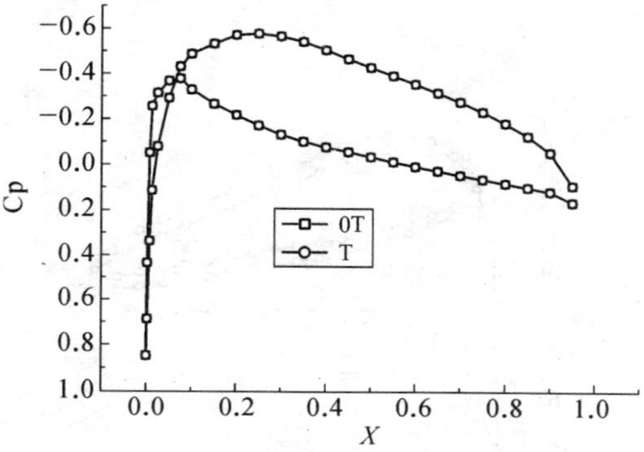

图 4.15 整周期压力对比

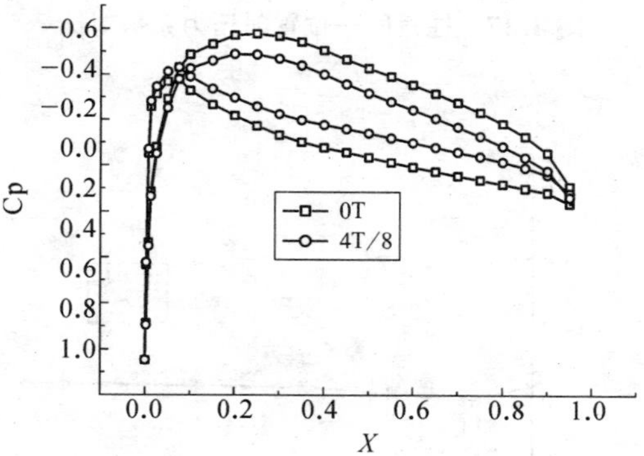

图 4.16 半周期压力系数对比

返同一个位置时的压力系数是相差比较大的，这一点同样可以从图 4.17 上看出。从尾涡线图 4.18 上同样可以看出我们的计算得到了周期性较好的结果。图 4.19 和图 4.20 都表明往返同一位置的尾涡线是有很大不同的。图 4.21 给出平移之后的五个时刻的尾涡线。图 4.22 给出了一个周期前后的 Mach 数对比，同样表明周期性满足得很好。

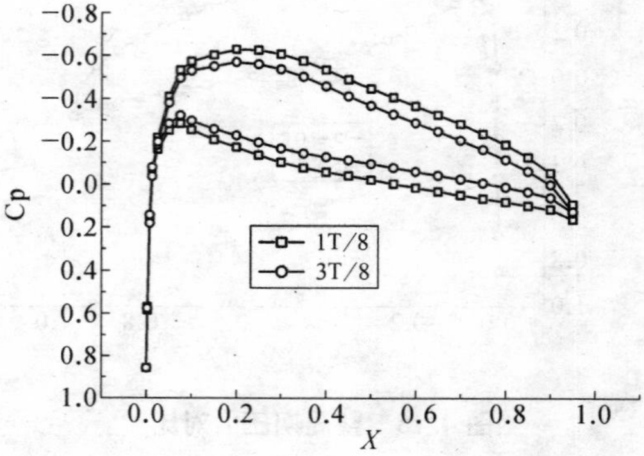

图 4.17　往返同一位置时压力系数对比

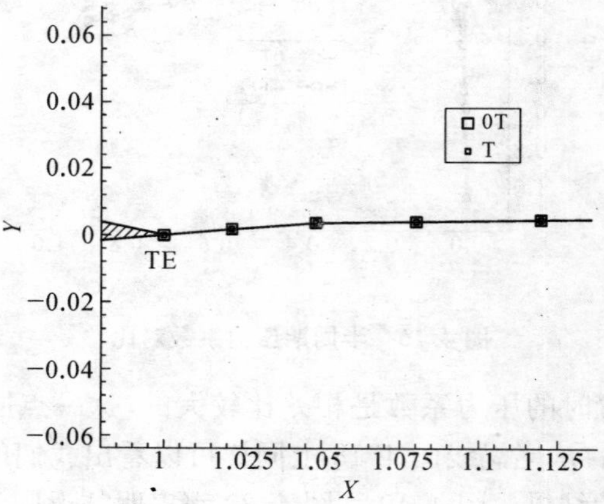

图 4.18　振荡一周尾涡线对比

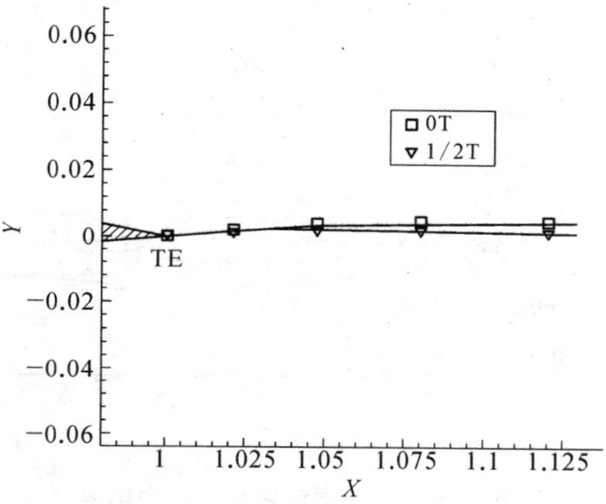

图 4.19　振荡半个周期尾涡线对比

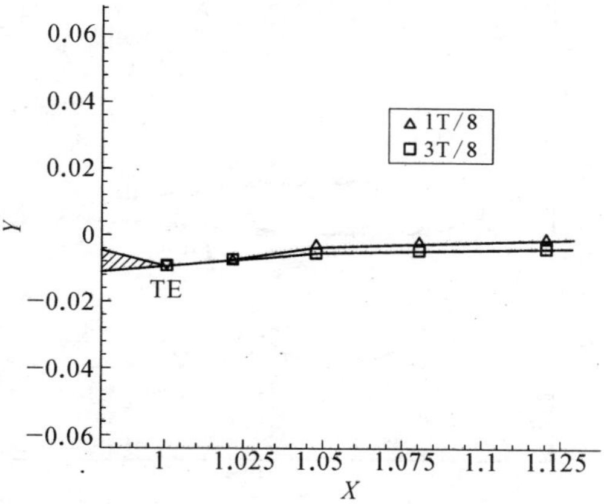

图 4.20　振荡往返同一非平衡位置时的尾涡线

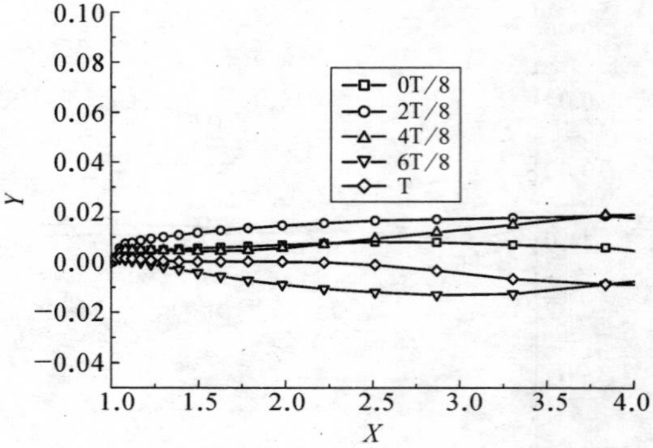

图 4.21　几条尾涡线平移后的结果

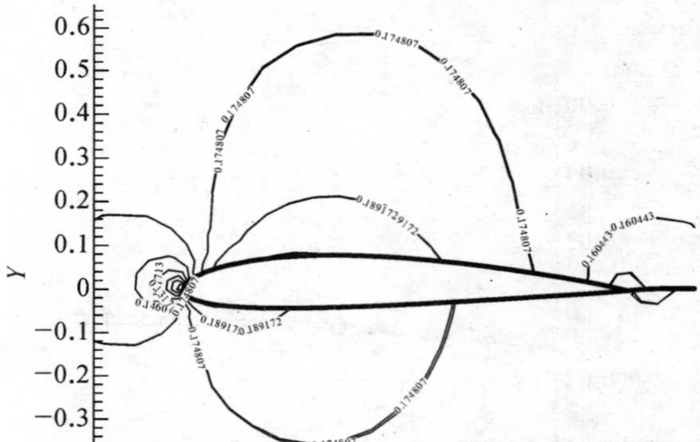

图 4.22　一个周期开始结束时的 Mach 数对比

4.2 非定常反命题求解

前一章我们已经得到了振荡翼型非定常绕流 $\mathrm{I_B}$ 型反命题的变分原理：该命题的解使下列泛函 J_1 取驻值：$\delta J_1 = 0$。其中 ϕ、A_2、A_s 和 A_f 各自独立变分，

$$J_1(\phi, A_2, A_s, A_f) = \frac{1}{\gamma}\iiint_{\Omega}\left\{1-(\gamma-1)\left[\frac{\partial\phi}{\partial t}+\frac{(\nabla\phi)^2}{2}\right]\right\}^{\frac{\gamma}{\gamma-1}}\mathrm{d}x\mathrm{d}y\mathrm{d}t + L \quad (4.9)$$

上式中

$$L = \iint_{A_1}(q_n)_{\mathrm{pr}}\phi\,\mathrm{d}s\mathrm{d}t + L_b + L_T$$

$$L_b = \frac{1}{\gamma}\int_{A_2}\left\{y_0\int_{t_{n-1}}^{t_n}[\mathring{p}+\delta(t-t_i)(p_{t_i}-\mathring{p})]\mathring{E}\mathrm{d}t\right\}\mathrm{d}x_0$$

$$L_T = \iint_{t_n}\mathring{\rho}\phi\,\mathrm{d}x\mathrm{d}y - \frac{1}{2}\iint_{t_{n-1}}[\rho_{\mathrm{pr}}\phi+(\phi-\phi_{\mathrm{pr}})\rho]\mathrm{d}x\mathrm{d}y$$

$$E = \frac{\sqrt{1+\left(\dfrac{\mathrm{d}y_0}{\mathrm{d}x_0}\right)^2}}{\sqrt{1+\left(\dfrac{\mathrm{d}y_0}{\mathrm{d}x_0}\right)^2+\left[(x_0-x_1)+\dfrac{\mathrm{d}y_0}{\mathrm{d}x_0}(y_0-y_1)\right]^2\left(\dfrac{\mathrm{d}\theta}{\mathrm{d}t}\right)^2}}$$

4.2.1 泛函的变域变分全时空有限元离散

下面我们用前面的全时空有限元法来离散反设计所用公式，求解 a（或势函数）和尾涡线（面）的方法和前面的完全类似，在此不作重复。由于我们最后所要求的是翼型的形状，所以上述泛函对 $\mathrm{A_2}$ 的变分是对翼型静止时的 y_0 坐标进行的，也就是说，离散后泛函求导也应

该对静止时 y_0 坐标求导。为此，我们必须用链式求导法则，即先对翼型的瞬时坐标求导，再对翼型的 y_0 坐标求导。翼型的瞬时坐标中 x，y 都与 y_0 有关，所以必须对他们都求导。由于这里是 $\mathrm{I_B}$ 型反命题，给定是目标分布是某时刻的压力分布，根据泛函式(4.9)我们知道，当 $t \neq t_i$ 时，由变域变分自然得到 $\left.\dfrac{\partial J_{1e}}{\partial y_{0i}}\right|_{t\neq t_i} = 0$，从而只当 $t = t_i$ 时有单元内的偏导数

$$
\begin{aligned}
\frac{\partial J_{1e}}{\partial y_{0i}} = & -\iiint_{\Omega}\rho\left[\frac{\partial J_3^{-1}}{\partial y_{t_n i}}\nabla'\phi + \frac{\partial J_1^{-1}}{\partial y_{t_n i}}\nabla'\phi \cdot J_1^{-1}\nabla'\phi + \right. \\
& \left.\frac{\partial J_2^{-1}}{\partial y_{t_n i}}\nabla'\phi \cdot J_2^{-1}\nabla'\phi\right]\frac{\partial y_{t_n i}}{\partial y_{0i}}\delta(t-t_i)\mid J\mid \mathrm{d}\xi\mathrm{d}\eta\mathrm{d}\zeta - \\
& \iiint_{\Omega}\rho\left[\frac{\partial J_3^{-1}}{\partial x_{t_n i}}\nabla'\phi + \frac{\partial J_1^{-1}}{\partial x_{t_n i}}\nabla'\phi \cdot J_1^{-1}\nabla'\phi + \right. \\
& \left.\frac{\partial J_2^{-1}}{\partial x_{t_n i}}\nabla'\phi \cdot J_2^{-1}\nabla'\phi\right]\frac{\partial x_{t_n i}}{\partial y_{0i}}\delta(t-t_i)\mid J\mid \mathrm{d}\xi\mathrm{d}\eta\mathrm{d}\zeta + \\
& \iiint_{\Omega}\frac{p}{\gamma}\cdot\frac{\partial\mid J\mid}{\partial x_{t_n i}}\frac{\partial x_{t_n i}}{\partial y_{0i}}\delta(t-t_i)\mathrm{d}\xi\mathrm{d}\eta\mathrm{d}\zeta + \\
& \iiint_{\Omega}\frac{p}{\gamma}\cdot\frac{\partial\mid J\mid}{\partial y_{t_n i}}\frac{\partial y_{t_n i}}{\partial y_{0i}}\delta(t-t_i)\mathrm{d}\xi\mathrm{d}\eta\mathrm{d}\zeta + \\
& \frac{1}{\gamma}\int_{A_2}\left\{N_i\int_{t_{n-1}}^{t_n}\delta(t-t_i)\,p_{t_i}\,\mathring{E}\mathrm{d}t\right\}\frac{\partial x_0}{\partial\xi}\mathrm{d}\xi \qquad (4.10)
\end{aligned}
$$

上式中的 N_i（对 8 结点 x、y、t 单元而言 $i=1$、2）为一维空间插值函数，表示为 $N_1 = \dfrac{1-\xi}{2}$，$N_2 = \dfrac{1+\xi}{2}$。$\mathring{E}$ 中的 x_0、y_0 都需要写成求和约定形式，即 $x_0 = N_k x_{0k}$、$y_0 = N_k y_{0k}$（对 k 求和）。$\dfrac{\mathrm{d}y_0}{\mathrm{d}x_0}$ 也应该写成

$\frac{\mathrm{d}y_0}{\mathrm{d}x_0}=N_k\frac{\mathrm{d}y_0}{\mathrm{d}x_0}\bigg|_k$ 的形式，当然为了计算简化，也可以直接用差分表示。

求得了各个时刻的 $\frac{\partial J_{1e}}{\partial y_{0i}}$，还需要求出它们的合成。合成后可以用前面采用的伪非定常的方式（参考式(3.26)和(3.27)，以及本章前面求尾涡的方式）求翼型的坐标改变量。改变翼型后重新生成网格再进行下面的迭代计算。由于这里的非线性很强，必须把伪非定常的时间步取得比较小，这里取为 0.2。

4.2.2 求解实例

我们这里只做了一个具有代表性的算例，是用对称的 NACA0012 翼型来设计非对称的 NACA2412 翼型。计算给定的分布是振荡一个周期中间时刻的压力分布，给定的参数就是正问题算例中例二计算得到的参数。设计总周期数定为 80，每三次正计算做一次反计算。

翼型坐标改变的收敛曲线如图 4.23 所示，可以看出当前的计算方法还是比较有效的。设计得到的翼型和目标翼型以及初始翼型的对比见图 4.24。可以由图上看出，设计翼型和目标翼型的主要区别在前后缘附近，这是因为我们的网格还比较稀疏，很难准确模拟前后缘的情况。但从图中我们还是可以看到设计翼型和目标翼型还是吻合得很不错的。设计所得翼型的一个周期中间时刻的压力系数分布和目标压力系数分布如图 4.25。可以看出设计翼型的压力系数分布和目标的分布很接近，主要区别还是在翼型前后缘附近。一个周期开始时目标翼型与设计翼型的压力系数对比见图 4.26，两翼型在八分之一周期的压力系数对比见图 4.27。从这些图上我们都可以看出，我们的设计翼型与目标翼型是比较接近的，证明我们的方法是可以用于非定常的 $\mathrm{I_B}$ 型反命题的求解的。同样，在翼型前后缘附近设计翼型和目标翼型相差比较大。图 4.28 是设计翼型和目标翼型的尾流线对比，可以看出两者的差别有所放大，这很可能是由于翼型后缘计算得不是很正确造成的。

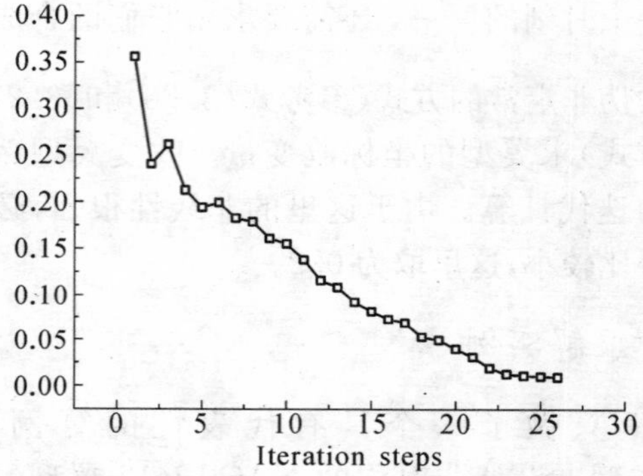

图 4.23　翼型坐标改变收敛曲线

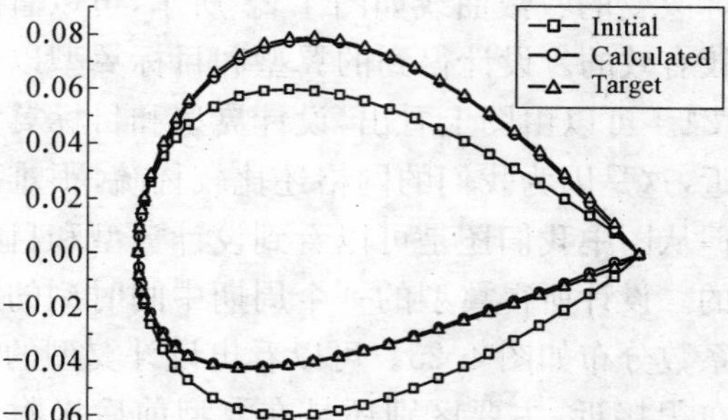

图 4.24　初始翼型、设计翼型和目标翼型对比

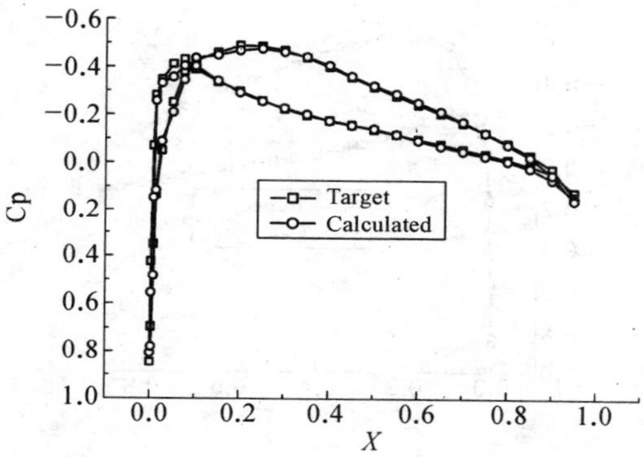

图 4.25　设计翼型的压力系数与给定压力系数对比

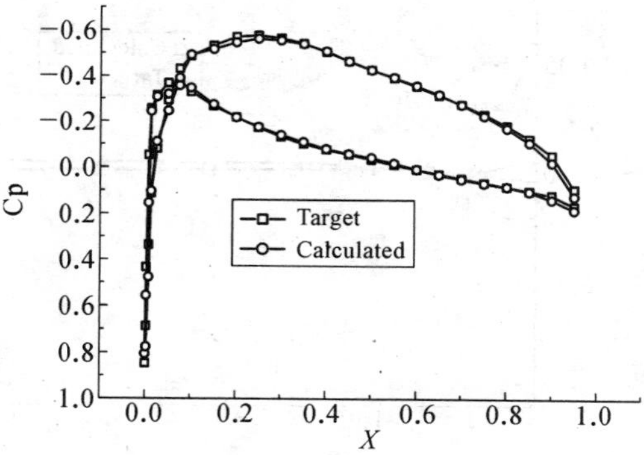

图 4.26　设计和目标翼型一个周期开始时压力系数对比

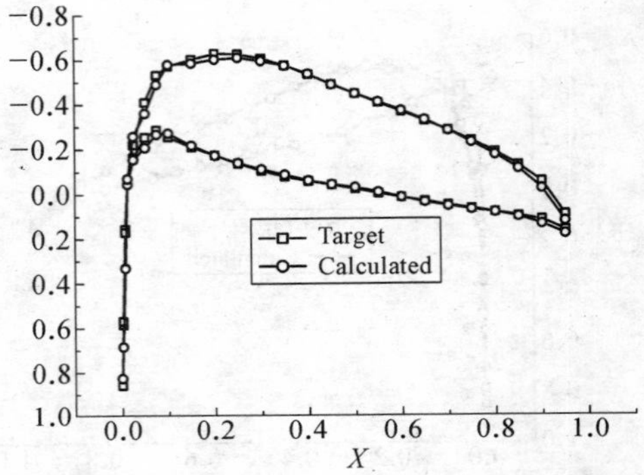

图 4.27　设计和目标翼型 1T/8 时的压力系数对比

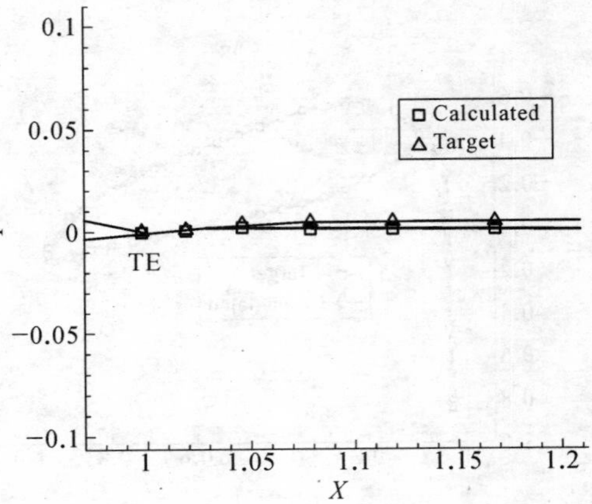

图 4.28　设计翼型和目标翼型的尾流线比较

4.3 小结

本章首先对翼型振荡的非定常正问题进行了变域变分时空有限元求解。通过对一个对称翼型和一个非对称翼型的计算表明,我们的方法能在计算过程中就捕捉到尾流线的位置,这是其他多种方法所不具备的。对该问题,我们根据多次计算得到了一些重要的经验系数。计算过程中还发现,Giesing-Maskell 所认为的后缘流动模型是正确的,翼型出流确实是平行于后缘上或下表面的。但由于计算尾流线的过程是一强非线性的过程,而且计算区域是变化的,属动网格问题,所以计算时间比较长。

对非定常反问题的求解,我们只做了一个有代表性的算例,是用 NACA0012 翼型来设计非对称的 NACA2412 翼型。得到的翼型和目标翼型主要是在前后缘附近存在一些误差,但总的说来还是模拟得很成功的,证明了我们的方法的正确性。经过改进,该方法还可能用于多工况的反设计中去,但计算量将是相当庞大的,在工程中应用还有很大难度。

第五章 结论与展望

5.1 结论

本文针对翼型的非定常可压缩流动的正问题和反问题进行了相关研究。具体讨论了本文的选题;翼型的非定常流动的正、反问题的合理提法,变分泛函的提出;新型全时空有限元及定常问题的时间相关解法;翼型的非定常流动正、反问题的全时空有限元求解等。得到了以下一些有意义的结论。

(1) 鉴于叶轮机向高速、高负荷方向发展,常规的设计方法已经不能满足设计需要,本文一开始就对叶轮机基于流场的设计方法研究进展进行了较为详细的综述,分析了各种现有方法的优缺点,提出进行非定常设计、多工况点设计的必要性。

(2) 由于有限元计算中梯度的计算精度对于原始变量的精度有决定性作用。本文对几种梯度计算方法进行了数值计算比较。得出采用算术平均的方法得到的梯度值能更真实的满足多种计算情况下的精度要求。

(3) 对定常可压缩流动下的翼型反问题采用了一种新的推导方式,即在翼型变分时限制其在 y_0 方向变分,使之可以用在非定常反问题变分原理推导过程中。通过计算得出采用变域变分有限元能较快地达到收敛,并给出了伪非定常时间步长在 0.5~1.5 之间。计算中还发现,我们的方法即使在初始翼型和目标翼型相差很大的情况下都能得到与目标翼型符合得很好的结果。

(4) 提出了一种计算非定常变域变分的新型简单全时空有限元插值模式,并成功地求解了一维非定常问题。对二维定常问题进行了非定常化的数值求解，也得到了相当满意的结果，充分说明我们提出的方法完全能进行非定常的变分计算。由推导过程可知该方法还可运用于高次多维情况，也为下一步的振荡翼型的非定常设计打下了基础。

(5) 推导得出了翼型的非定常正问题以及单工况的反设计的变分原理，考虑了多种边界条件的自然化处理。在反设计的变分原理中，采用了没有简化的振荡公式，使得文中的变分原理应用范围更广泛，人们更容易理解。在推导时保证了变分方向的唯一性。还推导得到了非定常单工况的反设计的广义、亚广义以及广义变分原理的普遍形式。

(6) 对振荡翼型非定常绕流正问题进行了变域变分求解。在离散时，采用了新型的全时空有限元法，并考虑了非定常 Kutta 条件的实施过程。数值计算表明，振荡翼型绕流的尾缘出流是与翼型尾缘上或下表面平行的。这与 Giesing-Maskell 模型吻合得相当好，也进一步证明了 Giesing-Maskell 模型的正确性。计算结果还表明，对称翼型在平均攻角为零的情况下处于零攻角位置时翼型表面是有升力的；往返同一位置时翼型表面同一点的压力系数是不同的。

(7) 对振荡翼型非定常反问题进行了变域变分求解。对当前提出的变域变分原理用新型全时空有限元进行了离散，得到了单元内的反设计公式。通过对一非对称翼型的反设计求解，证明我们的方法用于反设计是可行的，但计算时间相对较长。

5.2 展望

统观本文的内容，发现还可以在以下几个方面开展工作。

(1) 由于实际叶轮机是直列或环列叶栅，所以要推广使用还必须把该方法用于叶栅的可压缩、不可压缩的正反问题求解中去。

(2) 考虑到叶轮机一般都是工作在多个工况下的，很有必要进行翼型、叶栅的定常多工况设计。

(3) 目前的方法由于公式中采用的是势函数求解公式,所以从原理上严格地讲还不能考虑带激波的情况。而且,一般的流动都是有黏性的,不考虑黏性的计算与实际的结果有较大的差距。另外,建立黏性流动的控制变分原理目前来说难度很大。因此,下一步的一个重要方向就是寻求建立黏性流动的变分原理,并考虑带激波的情况。

(4) 上面的方法在计算时间上是很浪费的,做一个翼型的非定常反设计在 PIV1.7G 上需要几天时间。所以在下一步的工作中要寻求更省机时、更有工程实用意义的数值计算方法。

(5) 采用高阶的全时空有限元插值模式来进行计算。由于目前我们采用的是一阶的离散,所以精度不是很高。

(6) 考虑三维情况下叶轮机叶片的非定常和多工况反设计。在前面工作的基础上,开展这方面的工作就能促进本方法用在实际的工程中。

(7) 推广应用已有的理论和方法。这才是最终的目的,也是容易被忽视的。

参 考 文 献

[1] 刘高联. 叶轮机气动力学新一代反命题和优化设计的研究. 国家自然科学基金申请书，1999

[2] Liu G. L. Advances in research on inverse and hybrid problems of turbomachinery aerothermodynamics in China. Inverse Problems in Engineering，1996，Vol. 2：1～27

[3] 缪国平，刘应中等. 船舶流体力学中若干逆问题的研究. 力学进展，1996，26(4)：493～499

[4] Dang，T. Damle，S. Qiu，X. Progress with 3D inverse method for turbomachine blade design. American Society of Mechanical Engineering，Petroleum Division，1998

[5] Labrujere T. E.，Slooff，J. W. Computational methods for the aerodynamic design of aircraft components. Annual Rev. Fluid Mech.，1993，Vol. 25：183～214

[6] Yiu，K. F. C. Computational methods for aerodynamic shape design，Math. Comput. Modeling，1994，Vol. 20，No. 12：3～29

[7] Dulikravich，G. S. Aerodynamic shape design and optimization：status and trends. J. of Aircraft，1992，Vol. 29，No. 6：1020～1026

[8] Malone，J. B.，Vodyak，J. Inverse aerodynamic design method for aircraft components，J. of Aircraft. 1987，Vol. 24，No. 1：8～9

[9] Goldberg，D. E. Genetic algorithms：In search，optimization and machine learning，Addison-Wesley Publishing Company，Inc. Reading，MA，1989

[10] Holland, J. H. Adaptation in natural and artificial system, The MIT Press, Cambridge, MA, 1992

[11] Kirkpatrick, S., Gelatt, C. D., and Vecchi, M. P. Optimization by simulated annealing. Science, Vol. 220, No. 4598, 1983: 671～680

[12] Shelton, M. L. et, al. Optimization of a transonic turbine airfoil using artificial intelligence, CFD cascade testing. ASME paper 1993

[13] Rai, M. M., and Madavan, N. K. Aerodynamic design using neural networks. AIAA J. 2000, Vol. 38, No. 1: 173～182

[14] Lee K. D. Eyi S. Aerodynamic design via optimization. J. of Aircraft, 1992, Vol. 29, No. 6: 1012～1019

[15] Schmidt, E. Computation of supercritical compressor and turbine cascades with a design method for transonic flows. J. of Engrg and Power, 1980, Vol. 102: 68～74

[16] Wang Z. M. Inverse design calculation for transonic cascades. ASME Paper 85-GT-6

[17] 蔡立,李椿萱. 求解流体力学反问题的数学方法及应用. 空气动力学学报,1997,15(2): 198～205

[18] Takanashi, S. An iterative procedure for 3D transonic wind design by the integral equation method. AIAA paper 84～2155, 1984

[19] 赵小虎,阎超. 基于气动数值模拟的翼型反设计方法. 航空学报,1997,18(5): 648～651

[20] Quagliarella, D. Vicini, A. Viscous single and multicomponent airfoil design with genetic algorithms. Finite Elements in Analysis and Design, 2001, Vol. 37: 365～380

[21] Tiow W. T. Yiu KFC, Zangeneh, M. Application of simulated annealing to inverse design of transonic turbomachinery cascades.

Proceedings of the institution of mechanical engineers：Part A：J. Power & Energy，2002，Vol. 216：59～73
[22] Soemarwoto B. I. Labrujere，T. E. Airfoil design and optimization methods：recent progress at NLR. Int. J. Numer. Meth. Fluids，1999，Vol. 30：217～228
[23] Giles，M. and Drela，M. 2D transonic aerodynamic design method. AIAA J. 1987，Vol. 25，No. 9：1199～1206
[24] Zangeneh，M. Inviscid-viscous interaction method for 3D inverse design of centrifugal impellers. ASME paper 93 - GT - 103
[25] 王正明. 叶栅黏性流动的反问题解. 工程热物理学报，1989，10(4)：
[26] 王正明等. 叶栅全三维黏性反问题的数值解. 1998，19(5)：571～575
[27] Pierret，S. Van den Braembussche，R. A. Turbomachinery blade design using a NS solver and artificial neural network. J. Turbomachinery，1999，Vol. 121，No. 4：326～332
[28] Rhie，C. M. Gleixner，A. J. et al. Developments and applications of a multistage NS flow solver，Part 2：Application to high pressure compressor design. J. turbomachinery，1998，Vol. 120，No. 4：215～223
[29] Jameson，A. Schmidt，W.，Turkel，E. Numerical solutions of the Euler equations by finite volume methods using Runge-Kutta time stepping schemes. AIAA Paper - 81 - 1259，1981
[30] Jameson，A. Optimum aerodynamic design using control theory. CFD Review，1995：495～528
[31] 周恒. 新世纪对流体力学提出的要求. 自然科学进展，2000，10(6)：491～494
[32] Lagoudas D. C. Strelec，J. K. Yen，J. et al. Intelligent design optimization of a shape memory alloy actuated reconfigurable wing. Proceedings of SPIE，2000：338～348

[33] Kao. P. J. Parthasarathy, V. N. et al. Coupled aerodynamic-struactural shape optimal design of engine blades. Collection of Technical Papers - Proceedings of the AIAA/ASME/ASCE/AHS/ASC, Structures, Structural dynamics, and Matherials Conference 3, 1994 AIAA: 1317～1323

[34] Reuther, J., Jameson, A. Aerodynamic shape optimization of wing and wing-body configurations using control theory. AIAA paper 95-0123, 1995

[35] Hicks, R. M. Vanderplaats, G. N. Murman, E. M. et al. Airfoil section drag reduces at transonic speeds by numerical optimization. Soc. Automot. Eng. Paper, 760477, 1976

[36] Aidala P. V. Davis. W. H. Mason. W. H. Smart aerodynamic optimization. AIAA paper 83-1863, 1983

[37] Shiau, T. N., Chang, S. J. Optimization of rotating blades with dynamic-behavior constraints. J. of Aerospace Engineering. 1991, Vol. 4, No. 2: 127～144

[38] Huang, C. H., Hsing, T. Y. Inverse design problem of estimating optimal shape of cooling passages in turbine blades. Int. J. of Heat and Mass Transfer, 1999, Vol. 42, No. 23: 4307～4319

[39] Cheu, T. C., Bo, P. W. Design optimization of gas turbine blades with geometry and natural frequency contraints. Computers and Structures, 1989, Vol. 32, No. 1: 113～117

[40] Chattopadhyay, A., McCarthy, T. R., Pagaldipti, N. Multilevel decomposition procedure for efficient design optimization of helicopter rotor. AIAA J. 1995, Vol. 33, No. 4: 223～230

[41] Dulikravich, G. S. Proceedings of 3rd Int. Conf. on Inverse Design Concepts and Optimization in Engineering Sciences. ICIDES, Washinton, DC. Oct. 23-25, 1991

[42] Quagliarella, D. , Cioppa, A. D. Genetic algorithms applied to the aerodynamic design of transonic airfoils. In Proceedings of the 12th AIAA Applied Aerodynamics Conference, Colorado Springs, CO, USA, Jun, 1994, AIAA-94-1896-CP: 686～693

[43] Tong, S. S. , Powell, D. , Skolnick, M. Engeneous: domain independent, machine learning for design optimization. In Proceedings fo the 3rd Int. Conf. on Genetic Algorithms. (J. D. Schaffer Edit), pp. 151～159, M. Kauffmann Publishers, 1989

[44] Mossetti, G. , Poloni, C. Aerodynamic shape optimization by means of a genetic algorithm, Proceedings of the 5th Int. Symp. on Computational Fluid Dynamics, Sendai, Japan, pp. 279～284, 1993

[45] Fan H. Y. , Xi G. & Wang S. J. A dual fitness function GA and application in aerodynamic inverse design, Inverse Problems in Engrg. , 2000, Vol. 4: 325～341

[46] Doorly, D. J. , Peiro, J. , Oesterle, J. P. Optimization of aerodynamic and coupled aerodynamic-structural design using parallel genetic algorithms. In Proceedings of the 6th AIAA/NASA/USAF Multidiscipliary Analysis and Optimization Symp. Sept. 4-6, Seattle, WA, USA, 1996, AIAA Paper 96-4027, 1996

[47] Garabedian P. R. Mcfadden, G. Design of supercritical sweep wings, AIAA J. 1982, Vol. 20. No. 3

[48] Hyoung-Jin Kim Oh-Hyun Rho. Dual-point design of transonic airfoils using the hybrid inverse optimization method. J. Aircraft, 1997, Vol. 34, No. 5: 612～618

[49] Tapia, F. Sankar, L. N. Schrage, D. P. An inverse aerodynamic design method for rotor blades. J of the American Helicopter

Society. 1997, Vol. 42: 321～326

[50] Angrand, F. Methode numeriques pour des problemes de conception optimale en aerodynamique. These de 3eme cycle. L'universite Pierre et Marie Curie. Paris. 1980.

[51] Beux, F. Dervieux, A. Exact-gradient shape optimization of a 2D Euler flow. INRIA contr. Brite/Euram proj. 1082, part 1, 1991.

[52] Gu C. G. et al., Blade design of compressors by optimal control (I) & (II), ASME J. Turbomachinery, 1987, Vol. 109: 99～108

[53] Frank, P. D. Shubin, G. R. A comparison of optimization-based approaches for a model computational aerodynamics desigh problem. BCS - ECA - TR - 136, 1990

[54] Van Egmond, J. A. Numerical optimization of target pressure distributions for subsonic and transonic airfoil design. AGARD 1989, Pap. 16

[55] Obayashi, S. Takanashi. S. Genetic optimization of target pressure distributions for inverse design methods. AIAA J. 1996, Vol. 34, No. 5: 881～886

[56] Stanitz, J. D. Design of 2D channels with prescribed velocity distribution along the channel walls. NACA Report 1115, 1953

[57] Bauer F. Garabedian, P. Korn, D. Supercritical wing sections, Vol. 1, Springer-Verlag, New York, 1972

[58] Hassan, A. A. Sobieczky, H. Seebass, A. R. Subsonic airfoils with a given pressure distribution, AIAA J. 1984, Vol. 22, No. 9: 1185～1191

[59] Hassan, A. A. Dulikravich, G. S. A hodograph-based method for the design of shock-ree cascades. Inter. J. for Num.

Meth. In Fluids，1987，Vol. 7：197～213

[60] Garabedian P. R. Bledsoe，M. The method of complex characteristics for tranonic airfoil design，with an application to compressors，In Advances in Computational Transonics，Recent Advances in Numerical Method in Fluids，(Edited by W. G. Habashi)，Vol. 4，Pineridge Press，1985

[61] Lighthill，J. M. A new method of 2D aerodynamic design. ARC R&M 2112，1945

[62] Eppler R. Shen Y. T. Wing sections for hydrofoils，Part 1，J. Ship Research，1979，Vol. 23：209～217

[63] Selig，M. S. Maughmer，M. D. Multipoint inverse airfoil design method based on conformal mapping. AIAA J. 1992，Vol. 30，No. 5：1162～1170

[64] Selig，M. S.，Maughmer，M. D. Generalized multipoint inverse airfoil design. AIAA J. 1992，Vol. 30，No. 11：2618～2625

[65] Saeed，F.，Selig，M. S. Multipoint inverse airfoil design method for slot-suction airfoils. J. Aircraft. 1996，Vol. 33，No. 4：708～715

[66] Selig，M. S. Multipoint inverse design of an infinite cascade of airfoils. AIAA J. 1994，Vol. 32，No. 4：774～782

[67] Gopalarathnam，A.，Selig，M. S. Multipoint inverse method for multielement airfoil design. J. Aircraft，1998，Vol. 35，No. 3：398～404

[68] Gopalarathnam，A.，Selig，M. S. Low-speed natural-laminar-flow airfoils：Case study in inverse airfoil design. J. Aircraft，2001，Vol. 38，No. 1：57～63

[69] Henne，P. A. An inverse transonic wing design method. AIAA Paper 80－0330，1980

[70] Leonard，O.，Van den Braembussche，R. Design method for

subsonic and transonic cascade with prescribed Mach number distribution. J. Turbomachinery，1992，Vol. 114：553～560

[71] Demeulenaere，A.，Van den Braembussche，R. 3D inverse method for turbomachinery blading design. ASME Paper 96－GT－039，1996

[72] Demeulenaere，A.，Leonard，O.，Van den Braembussche，R. A 2D NS inverse solver for compressor and turbine blade design. Proceedings of the 2nd European Conference on Turbomachinery-Fluid Dynamics，and Thermodynamics，Antwerp，339～346

[73] Demeulenaere A. An Euler/Navier-Stokes inverse method for compressor and turbine blade design. Von Karman Inst. For Fluid Dynamics，Lecture Series 1997－05.

[74] Hawthorne，W. R. Tan，C. S. Wang，C. McCune，J. E. Theory of blade design for large deflections：Part 1：Two-dimensional cascades. J. Eng. For Gas Turbines and Power，1984，Vol. 106，No. 2：346～353

[75] Borges，J. E. A 3D inverse method for turbomachinery Part 1：Theory. J. Turbomachinery. 1990，Vol. 112，No. 7：346～353

[76] Zangeneh，M. A compressible 3D design method for radial and mixed flow turbomachinery blades. Int. J. Num. Methods in Fluids，1991，Vol. 13：599～624

[77] Zangeneh，M. Inviscid-Viscous Interaction method for 3D inverse design of centrifugal impellers. ASME J. Turbomachinery. 1994，Vol. 116：280～290

[78] Zangeneh，M. Goto，A. Takemura，T. Suppression of secondary flows in a mixed-flow pump impreller by application of three-dimensional inverse design method：Part 1 － Design

and numerical validation. Trans. ASME. J. Turbomachinery, 1996, Vol. 118. No. 7: 536～543

[79] Zangeneh, M. Inverse design of centrifugal compressor vaned diffusers in inlet shear flows. Trans. ASME J. Turbomachinery, 1996, Vol. 118, No. 4: 385～393

[80] Tjokroaminata, W. D., Tan, C. S., Hawthorne, W. R. A design study of radial inflow trubines with splitter blades in three-dimensional flow. Trans. ASME J. Turbomachinery, 1996, Vol. 118, No. 4: 353～361

[81] Liu, G. L. VP families for hybrid problems of blade-to-blade flow along axisymmetric streamsheet: A unified variable-domain approach (in Chinese), Acta, Aerodynamica Sinica, 1985, Vol. 3, No. 3: 24～32

[82] Liu, G. L. VPs for hybrid problems of transonic cascade flow along axisymmetric streamsheet: A unified variable-domain approach. Proceedings 4th Int. Symp. On Refined Flow Modeling and Turbulence Measurements, Sept. 1990, Wuhan, pp. 175～181

[83] Liu, G. L. A variable-domain variational theory using Clebsch variables for hybrid problems of 2 - D transonic rotational flow. Acta Mechanica, 1993, Vol. 99: 219～223

[84] Liu. G. L. A unified theory of hybrid problems for fully 3 - D incompressible rotor flow based on VPs with variable domain, ASME J. Engrg for GT and Power, 1986, Vol. 108, No. 2: 254～258

[85] Liu G. L. A variational theory of hybrid problems for fully 3 - D compressible rotor-flow: A unified variable-domain approach, Comput. Fluid Dynamics, G. d. v. Davis and C. Fletcher, (ed), Northholland, 1988: 473～480

[86] Liu G. L. Variational formulation of hybrid problems for fully 3 - D transonic flow with shocks in rotor, Proc. 3rd Int. Conf. on Inverse Design Concepts and Optimization in Engrg Sciences, Oct. 1991, Washinton, D. C. USA, pp. 337～346

[87] 刘高联. 二维机翼非定常气动力学反命题的变分理论. 空气动力学学报,1996,14(1):1～6

[88] Liu, G. L. A variational theory of inverse and hybrid problems for 2 - D unsteady free-surface flow around oscillating hydrofoils. Porc. National Symp. On Hydrodynamics' 93 Sept. 1993, Qinhuangdao, China, 632～637

[89] Liu, G. L. Unsteady inverse problem of type I_B for 2 - D transonic airfoil flow: a variational formulation. Proc. 3rd Int. Conf. Fluid Mech. Beijing Institute of Technology Press, Beijing, pp. 809～814

[90] Liu, G. L. A new generation of inverse shape design problem in aerodynamics and aerothermoelasticity: concepts, theory and methods. Aircraft Engineering and Aerospace Technology: An Int. J. 2000, Vol. 72, No. 4: 334～344

[91] Liu, G. L. A general variational theory of multipoint inverse design of 2 - D transonic cascades based on an artificial flow-oscillation model. Proc. 3rd Int. Conf. on Nonlinear Mech. Shanghai University Press, Shanghai, China, 502～508

[92] Liu, G. L. A systematic approach to the search and transformation for variational principles in fluid methanics with emphasis on inverse and hybrid problems. Experimental and Computational Aerothermodynamics of Internal Flow, Ed. By Chen N. X. Jiang, H. D. World Publ. Corp, Beijing, China, 1990: 128～135

[93] Finayson B. A. The Method of Weighted Residuals and VP.

Acad Press, 1972

[94] Poling, D. R. Telionis, D. P. The response of airfoils to periodic disturbances - the unsteady Kutta contions. AIAA J. 1986, Vol. 24, No. 2: 193～199

[95] Xu, JZ. Shock relations in turbomachines (in Chinese). Chinese J. Mech. Engrg, 1980, Vol. 16, No. 3: 60～69

[96] 郭加宏,刘高联. 机翼跨声速非定常绕流 I_A 型反命题变域变分有限元解,计算物理,2000,17:518～524

[97] Dulikravich, G. S., Sobieczky, H. Shockless design and analysis of transonic cascade shapes, AIAA J. 1982, Vol. 20, No. 11: 1572～1578

[98] Carey, G. F., Pan, T. T. Shock free redesign using finite elements, Communications in Applied Num. Methods. 1986, Vol. 2: 29～35

[99] Zhang, Z. Y., Yang, X. T. Laschka, B. Design of supercritical airfoil, J. Aircraft, 1988, Vol. 26, No. 6: 503～506

[100] Kubrynski, K. Design of 3 - D complex airplane configurations with specified pressure distribution via optimization. In Proceedings of the international conference on Inverse Design Concepts and Optimization in Engineering Sciences, (ICIDES-III), Washinton, DC, (Edited by G. S. Dulikravich), 1991

[101] Liu G. L. A vital innovation in Hamilton principle and its extension to initial-value problems Proc. 4th Intl. Conf. Nonlinear Mech. 2002, Shanghai Univ. Press, 90～97

[102] 刘高联. 耦合热弹性动力学的统一的变分原理族,力学学报,1999,31,(2):165～172

[103] Hughes T. J. R. Hulbert G. M. Space-time finite element methods for elastodynamics: Formulations and error

estimates, Comp Methods Appl. Meth. Eng., 1988, 66: 339～363

[104] 郭加宏,刘高联. 二维振荡机翼跨声速非定常绕流的变域变分有限元解. 空气动力学学报,1996,14 (4): 217～222

[105] 刘高联,陶毅,李孝伟等. 新型全时空有限元法. 待发表

[106] 陶毅,刘高联,李孝伟. 一维非定常流动一种新的全时空变分有限元插值模式. 第十七届全国水动力学研讨会暨第六届全国水动力学学术会议文集,2003,北京: 海洋出版社,179～185

[107] 李孝伟. 基于嵌套网格的全机带襟、副翼绕流的数值模拟. 西北工业大学博士论文,1999

[108] Oden, J. T. [1972]: Finite Elements of Nonlinear Continua, McGraw-Hill, New York

[109] Zienkiewicz, O. C., FRS, Taylor R. L. The Finite Element Method, 4nd Edt., McGraw-Hill, London, 1987

[110] T. J. Chung, Finite Element Analysis in Fluid Dynamics, McGram-Hill International Book Company, 1978 (有中译本)

[111] 王勖成,邵敏. 有限单元法基本原理和数值方法(第 2 版). 北京: 清华大学出版社,1997

[112] 朱伯芳. 有限单元法原理和应用. 北京: 中国水利水电出版社,1979

[113] Hirsch C., Num. Computation of Internal & External Flows, Vol. 2, 1990, John-Wiley, New York

[114] Sayanarayana B, Davis S. Experimental studies of unsteady trailing edge conditions. AIAA J., 1978, 16: 125～129

[115] Fleeter S. Trainling edge condtion for unsteady flows at high reduced frequency. AIAA J., 1980, 18(5): 497～503

[116] Maskell, E. C. On the Kutta-Joukowski condition in two-dimensional unsteady flow. Unpublished note. 1972 Roy. Aircraft Establishment, Farnborough

[117] Giesing，J. P. Vorticity and Kutta condition for unsteady multi-energy flows. Trans. ASME，J. Appl. Mech. 1969，91：608

[118] Basu，B. C. Hancock，G. J. The unsteady motion of a two-dimensional aerofoil in incompressible inviscid flow. J. Fluid Mech. 1978，78：159～178

[119] Batina，J. T. Unsteady Euler airfoil solutions using unstructured dynamic meshes. AIAA J. 1990，28(8)：1381～1388

[120] 陈池，刘高联. 基于变域变分有限元的翼型反设计，空气动力学报，2004，23(4)

攻读博士学位期间发表的学术论文及参与的科研项目

一、发表的学术论文

[1] 郑铭,陈池. 水锤数值计算的全特性曲线法,农业机械学报,2000,31：41～44

[2] 陈池,袁寿其等. 低比速离心泵叶轮内三维不可压湍流场计算,动力工程,2001,21：1346～1348

[3] 梁志勇,陈池. 微气泡对平板摩擦阻力影响的分析,上海大学学报,2002,8：267～272

[4] 刘高联,封卫兵,陈池. 流体力学广义变分原理临界变分状态的简便消除法,第十六届全国水动力学研讨会文集,2002：1～6

[5] 梁志勇,陈池. 用伪谱矩阵方法计算平板湍流边界层,船舶力学(EI 检索),2002,(1)：1～10

[6] 何有世,袁寿其,陈池. CFD 进展及其在离心泵叶轮内流计算中的应用,水泵技术,2002(3)：23～26

[7] 陈池,刘高联. 基于变域变分有限元的翼型反设计,空气动力学报,2004,23

[8] 陈池,陶毅,刘高联. 求解定常流动正命题的全时-空变分有限元法,力学进展,2004,25,(2)：157～162

[9] Chen Chi, Liu Gaolian. Aerodynamic Design of Airfoils Based on Variable-Domain Variational FEM, J. Shanghai University. Accepted

[10] Zhe-min Wu, Chi Chen, Gao-lian Liu. Multipoint inverse

shape design of airfoils based on varational principle. Aerocraft Engineering and Aerospace Technology，2004，Vol. 76，(4)：376～383

二、参与的科研项目

[1] 参与江苏省青年科技基金项目“离心泵水力设计进展及其内部流动三维湍流场数值模拟”，负责程序开发、数值计算、查新，并撰写论文、报告，负责答辩等。该项目已通过江苏省科委鉴定，并获得 2004 年江苏省科技进步一等奖，排名第二。

[2] 参与国家自然科学基金项目“离心泵内部的非定常流动研究”。

[3] 参与国家自然科学基金重点项目“叶轮机气动力学新一代反命题和优化设计的研究”，负责数值计算，项目阶段报告的整理等工作。

[4] 参与新型节能水泵“水波泵”的设计、试验。

致　谢

我的博士论文的完成，首先要对导师刘高联院士和李孝伟博士表示衷心的感谢。四年的博士研究生学习生活中，我从刘老师那里明白了如何有效地进行科研，如何做一个负责的、正直的科研人员。特别是导师追求科学真理的执着精神和一如既往的学术态度，给我留下了深刻印象。

我还要感谢课题组的老师黄典贵教授、封卫兵博士，以及博士生刘英学、吴兆春，硕士生陶毅、吴哲民、李红，他们对我的学习、工作和生活都提供了很多无私的帮助和建议。上海理工大学的姚征老师和西安交通大学的李开泰老师曾给我的计算工作提过良好建议，在此也感谢他们。

另外，邱翔、陈彤、章赛进、徐俊林、王泽晖、吉广伟、代数、袁邦兴、赵连霞、刘洁等等同学与我进行学习上的讨论，一起度过许多快乐的时光，在这里我也要向他们表示感谢。

我还要感谢力学所的所长郭兴明教授，所长助理董力耘博士，他们在学习、工作上给了我很大帮助。麦穗一老师的热情负责的态度是我以后工作的榜样，还感谢他对我们组内工作的极大支持。资料室的秦志强老师，她的热情和仔细令我感动。财务室的两位老师克勤克俭，热情为我们服务，为我顺利度过研究生生活提供了很多有益的帮助。谢谢了，力学所的老师们！

最后，我要感谢我的妻子郑丽云女士，她四年多的相夫教子任劳任怨，没有她的奉献，我不可能完成学业，不可能培养出我们的可爱的女儿。论文的完成与岳父母、父母、姐姐、弟弟的极大支持是分不开的，在此由衷地祝福他们身体健康，万事如意！

附　录　A

符号说明

	变量
$A_1 \cdots A_5$	三维区域的边界
c	当地声速,弦长
$\vec{e}$	可变边界位置矢量
E	函数
f	频率,函数
F, G	变域变分函数
H	焓
I	泛函
i	指标、x 向方向矢量(粗体)、虚数单位
J	泛函,Jacobi 矩阵
Jac	Jacobi 矩阵
j	指标、y 向方向矢量(粗体)
J	泛函
k	t 向方向矢量,指标,Riemman 不变量,折合频率 $k=\dfrac{\omega c}{2u_\infty}$
L	长度单位量纲
m	$=1/(r-1)$
n	迭代次数、法向矢量(粗体)
N	形状函数
p	压力

q	密流
R	圆的半径、Riemman 不变量
s	弧长
t	时间
T	时间周期
u, v, w	速度
x, y, t	空间、时间坐标
β, θ	角度
ε	阈值
γ	绝热指数
ξ, η, ζ	有限元坐标
ω	圆频率
ϕ	势函数
ρ	无量纲密度
Λ	无量纲速度
Ω	计算区域
∇	梯度
下标	
pr	给定的
∞	无穷
i, j, k	指针
e	单元
n	法向分量、时间层
τ	切向分量
s	激波
上标	
n	迭代次数
$*$	伴随
-1	逆
$'$	R^3空间中的量
o	限制变分

附　录　B

定常反命题的提法

二维可压缩势流的基本方程其相应的无量纲方程为

$$\nabla \cdot (\rho \vec{\Lambda}) = 0 \tag{B.1}$$

$$\nabla \phi = \vec{\Lambda} \tag{B.2}$$

$$(\gamma - 1)\left(\frac{\Lambda^2}{2}\right) + \frac{p}{\rho} = 1 \tag{B.3}$$

$$p = \rho^{\gamma} \tag{B.4}$$

上述公式中，ρ、Λ、ϕ、p、γ 分别为无量纲密度、速度、位势、压力和绝热指数。

这里选取的计算区域如图 B.1，相应的边界条件为

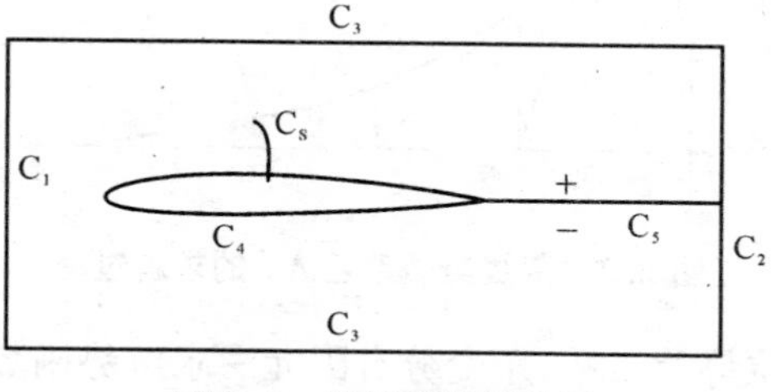

图 B.1　计算区域及边界

在进、出口边 C_1 和 C_2 上：

$$\left(\rho \frac{\partial \phi}{\partial n}\right)_{pr} = \rho_{\infty} \Lambda_{\infty} \cos \alpha。 \tag{B.5}$$

其中 ρ 为来流密度，Λ_∞ 为来流速度，α 为来流攻角。

在上、下边界 C_3 上：

$$\left(\rho\frac{\partial\phi}{\partial n}\right)_{\mathrm{pr}}=\rho_\infty\Lambda_\infty\sin\alpha。\tag{B. 6}$$

如果有比较小的激波，则在激波面 C_s 上：以 U_s 表示其法向分速，则有 Rankine-Hugoniot 激波条件[8]

$$\left.\begin{aligned}&[|\rho\Lambda_n|]=0\\&[|p|]/\gamma+\rho\Lambda_n[|\Lambda_n|]=0\\&[|\Lambda_\tau|]=0\\&[|H|]=(\gamma-1)\Lambda_n[|\Lambda_n|]\end{aligned}\right\}\tag{B. 7}$$

在割缝边界 C_5 上：周期条件 $\phi_+=\phi_-+\Delta\phi$，其中 $\Delta\phi$ 为绕翼型的势函数差（即绕翼型的环量）是通过 Kutta 条件迭代求解出来的。具体的过程是（参看图 B. 2）：

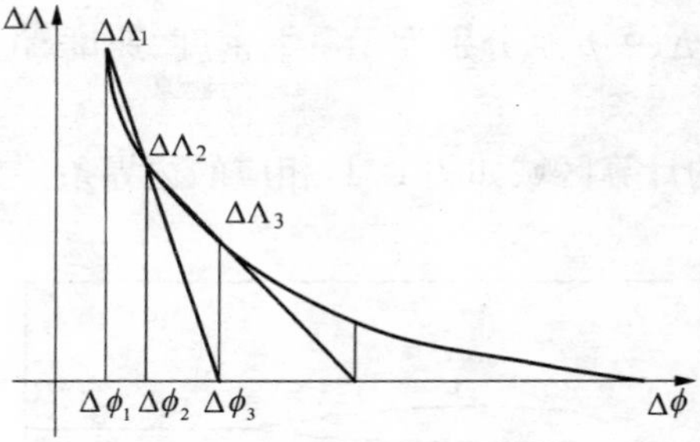

图 B. 2　割缝势函数差 $\Delta\phi$ 的求解过程

(1) 设一位势差 $\Delta\phi_1$，用变分有限元法求解势函数场。

(2) 求解速度场，得到翼型尾缘附近上下对应点的速度差 $\Delta\Lambda_1$。再以位势差 $\Delta\phi_2=1.5\Delta\phi_1$ 代入割缝边界条件，重新求解势函数场，再计算翼型上下的速度差 $\Delta\Lambda_2$。在这里我们选取的对应点为靠近尾缘的两点，它们分别在翼型的吸力面和压力面上，具体位置可以适当给定。

(3) 把上面计算得到的 $\Delta\phi_1$、$\Delta\Lambda_1$、$\Delta\phi_2$ 和 $\Delta\Lambda_2$ 代入下式：

$$\Delta\phi_3 = \Delta\phi_1 - (\Delta\phi_2 - \Delta\phi_1)\ X\ \Delta\Lambda_1/(\Delta\Lambda_2 - \Delta\Lambda_1)$$

(4) 求得校正的 $\Delta\phi_3$ 值，又可求解出新的流速场及尾缘处的速度差 $\Delta\Lambda_3$，如果它小于某个阈值或迭代进行了一定次数则停止迭代，否则把 $\Delta\phi_2$ 和 $\Delta\Lambda_2$ 分别赋给 $\Delta\phi_1$ 和 $\Delta\Lambda_1$，把 $\Delta\phi_3$ 赋给 $\Delta\phi_2$，再次求解速度场，得到 $\Delta\Lambda_2$，再回到步骤(3)。

尾缘点上：应满足 Kutta 条件，即要求尾缘点附近上下的速度相等。

由变分推导的系统性途径，我们可以得到二维可压缩势流的变分原理：二维可压缩势流的解将使：$\delta I' = 0$，其中

$$I'(\phi) = \frac{1}{\gamma}\iint\limits_{(\Omega)} \left\{1 - \frac{1}{2m}\left[\left(\frac{\partial\phi}{\partial x}\right)^2 + \left(\frac{\partial\phi}{\partial y}\right)^2\right]\right\}^{\gamma m} \mathrm{d}A + \int\limits_{(C_{1,2,3})} (q_n)_{\mathrm{pr}}\phi\,\mathrm{d}s \tag{B.8}$$

下面用变域变分来推导翼型反设计所用的变分泛函。我们这里的泛函其变域变分公式为

$$\delta I' = \iint\limits_{\Omega}\left(\frac{\partial F}{\partial\phi} - \nabla\cdot\vec{G}\right)\tilde{\delta}\phi\cdot\mathrm{d}A + \int\limits_{\partial\Omega}\{G_n\cdot\delta\phi + (F\vec{n} - G_n\nabla\phi)\delta\vec{e}\}\mathrm{d}s \tag{B.9}$$

这里的 F、G、G_n 的定义分别为

$$F = \frac{p}{\gamma} = \frac{\left\{1 - \frac{\gamma-1}{2}\left[\left(\frac{\partial\phi}{\partial x}\right)^2 + \left(\frac{\partial\phi}{\partial y}\right)^2\right]\right\}^{\frac{\gamma}{\gamma-1}}}{\gamma}, \tag{B.10}$$

$$\vec{G}=\frac{\partial F}{\partial\phi_x}\vec{i}+\frac{\partial F}{\partial\phi_y}\vec{j}=-\rho\vec{\Lambda},\tag{B.11}$$

$$G_n=\vec{G}\cdot\vec{n}。\tag{B.12}$$

对翼形表面 C_4 使用变域变分，有

$$\delta I=\iint_{\Omega}\nabla(\rho\Lambda)\delta\phi\,\mathrm{d}x\mathrm{d}y+\int_{C_4}\frac{\partial\phi}{\partial n}\delta\phi\,\mathrm{d}s+$$

$$\int_{C_4}\left(\frac{p}{\gamma}-G_n\frac{\partial\phi}{\partial n}\right)\delta e_n\mathrm{d}s-\int_{C_4}\frac{\partial\phi}{\partial\tau}\delta e_\tau\mathrm{d}s\tag{B.13}$$

为了计算方便，在此规定边界变分只是沿着 y 向进行，即有

$$\delta\vec{e}=\delta y\cdot\vec{j}\tag{B.14}$$

参看图 3.6，于是有

$$\delta\vec{e}\cdot\vec{n}\cdot\mathrm{d}s=\delta y\cdot\vec{j}\cdot\vec{n}\mathrm{d}s=-\delta y\mathrm{d}x,\delta e_\tau=0\tag{B.15}$$

又由 $\delta J=0$ 知在 C_4 上，应有：$\frac{\partial\phi}{\partial n}=0$，故得：

$$\delta I=\iint_{(A)}\nabla(\rho\Lambda)\delta\phi\,\mathrm{d}x\mathrm{d}y+\int_{C_4}\left(\frac{p}{\gamma}\right)y\delta e_n\mathrm{d}x\tag{B.16}$$

由上式可看出，只要在 I' 中增补一项 $\int_{C_4}\frac{p_{\mathrm{pr}}}{\gamma}y\mathrm{d}x$，即得该反问题的变分泛函

$$I'(\phi,Y)=\frac{1}{\gamma}\iint_{(\Omega)}\left\{1-\frac{1}{2m}\left[\left(\frac{\partial\phi}{\partial x}\right)^2+\left(\frac{\partial\phi}{\partial y}\right)^2\right]\right\}^{\gamma m}\mathrm{d}A+$$

$$\int_{(C_{1,2,3})}(q_n)_{\mathrm{pr}}\phi\,\mathrm{d}s+\int_{(C_4)}\frac{p_{\mathrm{pr}}}{\gamma}y\mathrm{d}x。\tag{B.17}$$

式中的 Y 表示翼型的所有可变的 y 坐标。